JUNGLE PATROL

Vic Hurley

HAS ALSO WRITTEN

SOUTHEAST OF ZAMBOANGA

"The value of Hurley's story lies in its accurate reporting. His book is easy reading."

—*The Saturday Review of Literature*

MEN IN SUN HELMETS

"A copious variety of stories here to suit the tastes of every man, and an admirable complementary volume to the author's previous work." —*The New York Sun*

SWISH OF THE KRIS

"—all vividly interesting."

—*The San Francisco Call-Bulletin*

". . . Hurley knows the Philippines and understands the islands far better than the average American who is drawn there for business reasons." —*The Washington Post*

". . . because he (the author) has known the Moros and sympathized with them, he has been able to synthesize all this into a thoroughly interesting and probably important book." —*New York Herald Tribune Books*

PUBLISHED BY E. P. DUTTON & CO., INC.

THIS 2011 EDITION OF *JUNGLE PATROL, THE STORY OF THE PHILIPPINE CONSTABULARY (1901 – 1936)* CONTAINS THE COMPLETE AND UNEDITED TEXT OF THE ORIGINAL 1938 EDITION.

Other Books by Vic Hurley:

- *Arrows Against Steel, the Story of the Bow and How It Forever Changed Warfare*
- *The Parthian*
- *Incidents in Asia*
- *England and Elsewhere*

Early Constabulary officers — Allen with arms folded

JUNGLE PATROL

THE STORY OF THE PHILIPPINE CONSTABULARY 1901-1936

By VIC HURLEY

With Original Photographic Illustrations

Maps by Christopher L. Harris
Based on Original Period Maps
Of Paul H. Kirk

CERBERUS BOOKS

2011

Library of Congress Information
LC Control No: 38032187; LC Classification: HV8255.A2 H8
Subjects: Police -- Philippines

ISBN: 13: 978-0-9834756-2-0; 10: 0-9834756-2-8

Cover Art, and Maps (Based on Original Period Maps of Paul H. Kirk)
by Christopher L. Harris
Foreword by K. W. Kersh

BISAC Classification:
HIS 048000 HISTORY / ASIA / SOUTHEAST ASIA
LAW 026000 LAW / CRIMINAL LAW / GENERAL
HIS 027180 HISTORY / MILITARY / SPECIAL FORCES

PRINTED IN THE USA AND OTHER COUNTRIES BY ARRANGEMENT

Cerberus Books (Imprint)
Cerberus Corp, Box 13048, Salem, Oregon, 97309

FOR JIM

Foreword

Vic Hurley was an Honorary Third Lieutenant in the Philippine Constabulary. After the success of his non-fiction Philippine Books, *Southeast of Zamboanga* and *Men in Sun Helmets* he produced this work in the belief that the world had overlooked the exciting and unique story of these brave and courageous officers and men.

At the end of the nineteenth century, following the Spanish-American War, American public opinion had turned against foreign military engagements. Although there were still more that 70,000 U.S. troops on duty in the islands, the repression of Philippine insurgencies and, later, banditry and piracy was determined to be unpalatable to the political panjandrums in Washington. Thus, in 1901 the Constabulary was initiated as a political expediency by the McKinley administration and was created by the U.S. Philippine Commission in Manila headed by William Howard Taft.

President McKinley, abandoning the Monroe Doctrine, made no effort to conceal the colonialistic aims of his administration during what was later called the 'Philippine-American War". It was conducted, at first, by U.S. troops against native Filipino military forces and subsequently by the Constabulary against loyalist insurgents, bandits, and pirates. McKinley said, "While we are conducting this war and until its conclusion, we will keep all we get; when the war is over, we must keep what we *want*."

The American political landscape of the time is beyond the scope of this book, but the influence of sugar barons should not be discounted. The American military and political habit of investing U.S. forces in cultures and environments for which they are notably unsuited, may have begun here, but has never subsequently ceased.

The U.S. military command in the Philippines was instructed to publicly report only positive news and its dispatches during this period were, to say the least, misleading. The U.S. Army was unsuited and unprepared to wage war in a jungle environment where every advantage lay with those who, at the outset, believed they were resisting yet another foreign occupier – having just rid themselves of the Spanish.

What the Commission envisioned was a para-military domestic police force composed of foreign officers (American and European) and native recruits, to be charged with suppressing insurgencies, and other anti-American and criminal strife – throughout the entire archipelago. The training of the native forces and the development of loyalty to be left with the 'foreign' officers – who were also to lead in the subsequent engagements.

The U.S. military was chary about arming the native Filipino 'policemen' fearing desertion and armed rebellion and intervened to prevent them from having weapons equal to the weapons of their foes.

While the British in India had had success with native forces led by English officers, the American Army Command in the Philippines was skeptical of the loyalty of such a native force, although it also had a Filipino corps – the Philippine Scouts. It did on occasion offer army help for Constabulary actions, but such help was never guaranteed and at some command levels, there was open disdain.

FOREWORD

Hurley, of course, narrates this colorful epic from the viewpoint of an American patriot, but his respect for the Filipinos – particularly their organized armies of resistance and the subsequent loyalist guerilla actions, is clear. Later, native opportunists of one stripe or another used the pretext of rebellion to savage foreign and native populations with banditry and piracy.

There needs to be special mention of the example of the tragic history of one of these Constabulary officers: Leonard Furlong. Of the many brave officers that led the Constabulary, he is acknowledged to be in the forefront of this doughty group. Many of the dispatches from the period feature Constabulary battles and engagements led by Furlong. They have an almost mythical content. Here is an example:

[Sultan Dimabara, his father, his two brothers and one other had assassinated an American Army soldier, Private McDonald. Lt. Furlong was ordered to apprehend the miscreants at Bugsan Village, Cotobato District, Moro Province, Mindanao Island.]

> "August 7, 1906. Word was received on July 9th describing the dwelling place of Sultan Dimabara. On the next night Lieut. Furlong with five Constabulary men, two scouts and four moros, the entire force having but six rifles, went to this dwelling place arriving there before daylight, July 11th. Two of the four taking the wrong trail near the place were chased back to the party by hostile moros.
>
> "Sultan Dimabara and his outlaw companions were called to come out of his house and surrender. Instead of doing so their people called to each other to come out with their weapons and fight. A hundred moros then surrounded Lieutenant Furlong's party, while those in the house repeated their religious oaths to die together.

FOREWORD

"The fighting then became general.

"Under these conditions Lieut. Furlong exhibited excellent judgment and exceptional gallantry.

"By the skillful handling of his men and his quickness in action Lieut. Furlong succeeded in killing two of the assassins, the father and brother of Dimabara, and four other hostile moros. He then extricated himself and his party and passed out of the place without injury to his force. (LETTER OF THE DISTRICT GOVERNOR OF COTOBATO. Dated August 7, 1906)

Although Furlong acted in accordance with the orders of his superior officers and the Commission, in 1909 he was charged with the wanton destruction of Filipino property and lives in the Buldung region. He, himself, requested a court-marshal and was eventually cleared of all charges. Sadly, he apparently concluded that his military career was over, and in 1911, took his own life, as had a surprising number of other officers.

The American Philippine policy under McKinley and his successors made the job of the Constabulary doubly difficult. On the one hand it was charged with the domestic responsibility for establishing peace in the islands by the removal of dissident individuals and movements, and, on the other, abiding by concessions made by regional American administrators in their attempts at unrealistic pacification. Many of these officials were openly hostile to the Constabulary.

Although the book ends with the Constabulary being merged with the nascent National Army of the Philippines in 1936, it was, in fact, re-established as a separate entity several times. As of this writing it is again a part of the Philippine National Police.

FOREWORD

This book was created by Hurley from the records of field dispatches and the personal recollections of many of the men who served, and officers who led the Constabulary. He had the good fortune to find and communicate with some of the legendary personalities of this epoch before their deaths: Cromwell Stacy, Cary Crockett, A. E. Hendrix, and Jesse Tiffany, among others, and he had access to other written records and memorabilia of the era, now rare or non-existent, as is disclosed in the Bibliography.

In preparing this re-issue we have made no changes to the original text. We have improved for clarity two of the Kirk maps that did not reproduce well in the original edition.

K.W.K
June, 2011

Contents

Jungle Patrol

CONTENTS

List of Illustrations

List of Illustrations

Introduction

THE men who pacified the Philippines were a strange breed. They were world-weary gentlemen and their job was the taming of jungle. They were a fighting breed, of an age when combat was hand-to-hand, rather than the impersonal hatred at ten miles that is the rule of warfare today. Yes, they *were* strange men—they were battlers who negated the cautious principles of warfare that actuate the modern scientific soldier. Apparently absent from their make-ups were the fundamental emotions of fatigue and doubt and terror.

To call them "Men without Fear" would be a trite and a possibly inaccurate over-enthusiasm. And yet the fiery combat records of the Philippine Constabulary seem to substantiate that heroic label. The flame-flecked years of the early 1900's in the Philippine Archipelago are without parallel in the military history of our nation. These jungle soldiers accepted odds as a necessary evil of the military system. The ghastly gurgle from the slit throats of their sentries was often the first warning of the rush of the bolomen; the soldiers who died on the dreadful patrols were buried where they fell. The wounded struggled on through the swamps to die in turn, or to come back with the honorable scars of conflict.

The campaigns of the Philippine Constabulary, the regular army, and the Philippine Scouts constituted America's first experiment in warfare in the jungle. In great measure, it was the bloody aftermath of the bloodless Spanish-American war that brought the United States into prominence as a world power. The Constabulary was a unique and successful application of the principle of employing native infantry, officered by white men, in the subjugation of their own tribesmen. Other nations had used the principle of recruiting native soldiers but the Constabulary developed a fundamental difference in the application of the force. This insular police unit fought the natives of a district with troops recruited in the same district.

They eliminated organized banditry in the Philippine Archipelago, and like terriers they pursued the scattered fragments of such bands until peace came to the Islands.

They accomplished this result with a loss, during that six-year period, 1901-1906 inclusive of 1,029 men. This casualty percentage is very high and is perhaps the best indication of the severity of the jungle campaigns. The Constabulary was never in excess of seven thousand men. They were consistently outnumbered as they conducted their punitive, offensive operations in the bush. Their foeman ambushed and slew from the shelter of that silent jungle.

Knowing the nature of their work, it would be easy to assume that the Constabulary was therefore a heavily armed force, with superior weapons to offset that disparity in numbers. Such was not the case. The *pulajans* of Samar were invariably better armed than was the opposing police force, which, until late in its career, was equipped with very deficient weapons. The massacre at Balangiga, by *pulajans,* of a company of the 9th Infantry, United States Army, in September 1901, provided the hillmen with more modern repeating rifles than were possessed by the Constabulary for some years.

In their beginning, the Constabulary was armed with single shot Remington shotguns, with an effective range of about one hundred yards. This arm was supplemented by the Colt single-action calibre .45 revolver. Against these short-range weapons, the *pulajans* brought to bear Mauser rifles of the Spanish army and the American Krag-Jorgensen. Later in its existence, the Constabulary was equipped with cast-off Springfield rifles, with black powder ammunition and a capacity of a single shot. This was the old calibre .45 army rifle. It was not until 1906, when the wars were dwindling to a close, that the Insular Police received repeating Krags and then only after the army had discarded the Krag in favor of the modern .30-06 Springfield.

During the fiercest of the campaigns, the Constabulary had time for but one shot before the melee became hand-to-hand. There was no time to reload and they had no bayonets to oppose the bolo rush. They fought with clubbed rifles, teeth, and fists. Their uniforms were nondescript; their commissary and medical divisions were practically non-existent and their reinforcements were nil. They occupied isolated stations—a few men with a Lieutenant, or fewer men under a Sergeant. They had no regimental organization, either then or later in their career. The company was the unit—more often, the platoon or the squad. They made long marches into the mountainous jungle in search of antagonists who fell back before the police advance until the moment came to strike.

The Constabulary lived off the jungle — on a diet of python and rat and fruit bat. They carried their wounded with them as long as was possible and then they retired to heal them to return to fight again. Their dead they buried in the sombre jungle that claimed them.

The Constabulary of this early day had no pension provision and no relief for the widows of the fallen. Their pay was painfully inadequate; it was not unusual for Lieutenants to spend pay vouchers for cigarettes for their men. A Brigadier-General, in the top spot, drew $3,000 annually. A Captain signed for $1,100. A Third Lieutenant contented himself with an annual award of $800. The private fought for $6 each month.

The marvel of it all was the splendid efficiency of these

military orphans. By sheer merit, they won the respect of the regular army and of the administration that created them. They performed a service that no massed troop movement of the regular army could possibly have contributed and they did it with an average expenditure of $250 per year per soldier, in comparison to the $1,000 spent each year on the regular army trooper in the Philippines.

In undertaking the preparation of this record of the Philippine Constabulary, I do so with but one regret. That, the inadequacy of my pen. To the officers who have made possible this account, I convey my most sincere thanks. I am proud of the confidence with which they have turned over to me the combat orders, diaries, and personal memoirs of this most neglected chapter of American military history.

VIC HURLEY

Seattle, Washington
August, 1938

JUNGLE PATROL

"To be outnumbered, always; to be outfought, never."

—*The Creed of the Jungle Police*

Chapter i

A SETTING AND A MOOD

"I navigated to the Islands of the Philippinas, hard on the coast of China; of which countery have I brought intelligence."

—Thomas Candish, 1588

THIRTY-NINE years ago, a young and powerful nation of the West turned aside from its principles of democracy and freedom for all and sought to impress its sovereignty upon the scattered peoples of a tropic archipelago.

The armed men who marched away to do this thing were young Americans; figures in slouched campaign hats and blue shirts, who were not impressed too greatly with the discipline of their military system. With levity and song they won a minor and almost bloodless war for the United States, and in so doing, they gained the wardenship of divers tribes of the Malays.

This war against a second-rate power that resulted in the capture of Manila from the Spaniards was the joke—the future pacification of the native tribesmen was to become the joker.

We must turn back to the lusty, careless days of '98 for a realization of the inexperience of the United States in the undertaking of any manner of sustained jungle warfare.

February 15—April 24, 1898. . . .

The *Maine* sinks; the white man's burden is taken up with enthusiasm; it is the glamour period, with a self-righteous tinge of altruism to color the scene. To the refrain of "There'll Be a Hot Time in the Old Town Tonight," the nation moved to war. It was a different kind of war—

and a good war. Even the pacifists realized that. Here was a war fit for the best of the American traditions—an extension of the Monroe Doctrine to the East, as it were, and waged, not for territorial acquisition, but as a big-brother defense of a little brown people who were cringing under the lash of Spain. The cartoons in the newspapers of 1898 were filled with beetle-browed Spaniards with unpleasant, leering features. It made fine reading in the daily journals.

But only the sugar men knew what the war was about.

MAY 25, 1898. . . .

The chartered steamers *City of Pekin, Australia,* and *City of Sydney* roll at anchor in San Francisco Bay. For a month the guns of the Spanish fleet in Manila have been stilled, and the battered hulks are gathering rust at the bottom of a tropic bay. A few of the better informed in America actually know where Manila is located, as Dewey rides at anchor there, surveying the city, with no landing force available to take possession. And so the docks at San Francisco are scurrying with martial activity. The first expeditionary force is embarking—America is off to take possession of an archipelago. But still the flavor is one of altruism.

Humor, pathos, ghastly mistake, and tragedy are all a part of this expeditionary force. The American authorities are ignorant of the fundamentals of tropic sanitation, and the troops are ill-clad for service on the equator. They wear heavy service uniforms, with blue shirts, and they carry overcoats across their arms! They are fortified from chill with heavy underwear.

Their combat essentials are ill-distributed about their persons. It is before the day of shoulder-packs; all equipment is slung from the hip. It is a galling weight, with canteen, haversack, and a double row of 100 cartridges.

Jungle Patrol

A SETTING AND A MOOD

On August 13, 1898, these troops participated in a battle before the moss-grown walls of Old Manila, and the result was the lowering of the flag of Castile and Navarre.

The taking of Manila was accomplished in a day. It was an historical incident, designed to save the face of Spain, whose defense was a feeble gesture. The winnings of the day were thirteen thousand Spaniards, an array of ancient cannon admirably suited for display in the public parks of the United States, and $900,000 of public money of the treasury of Manila.

With the capture of Manila, it was confidently believed that the war was over, and the martial mood changed to one of riotous revelry. Soldiers who had won a war were entitled to relaxation. And it was a land where señoritas smiled from shaded balconies. Three hundred saloons opened with startling rapidity, and under the stimulus of the warming cheer they dispensed, young troopers forgot all about Spain and began a series of all-American private wars. They also forgot all about the Filipinos, who were beginning to realize that the war had resulted in but a change of masters.

Turbulent Manila in '98: A regulation is provided that all troops be off the streets by seven o'clock in the evening. To enforce it, a sad 20th Infantry is detailed to the unpopular job of Military Police duty. A song is born, stigmatizing the 20th, the mere whistling of which was sufficient to participate a street brawl:

"The bridge of Spain,
Will groan with pain,
When the 20th goes to battle."

Distinct in the memories of the genuine old-timers who

served in Manila during this ribald aftermath of war is the "affair of the test tubes," and it concerned the 28th Infantry.

It happened in this wise:

All day long the men of the regiment had stood in a sultry, sulking line, awaiting their turn to yield blood specimens for examination by the medical force. Grumbles and protests rolled along the file as they sweated in the broiling sun, at the dubious pleasure of overworked physicians.

At last, after weary hours, the doctors were finished with their victims and the test tubes of blood specimens were racked in the infirmary, pending examination on the morrow. The men heaved a sigh of relief and filed back to their quarters. The record of their iniquities was on file and they need stand no longer in the steaming heat.

But that night, in the dark hours, come three genial drunks of the 28th Regiment, to break in the door of the infirmary with that curious lack of logic of the alcohol-saturated. Weaving their way along the uninspiring aisles, one of the number spies the long line of ranked test tubes. Jovially he reaches for one and pours the contents upon the closest of his inebriated companions.

The fight begins. . . .

Test tubes begin to fly, the contents spattering the walls as the earnest antagonists riot there, boiling with liquor. One of the combatants sobers suddenly as he catches sight of his face in a mirror. Horror-stricken, he rushes from the infirmary, moaning with terror and anguish, "I'm bleeding to death. They have killed me."

A Sergeant of the Guard bears down upon him, muttering sourly, "Not yet they haven't. You have that to look forward to after you have peeled potatoes for a million years. But they damn well *will* kill you when they learn they have to stand in line again tomorrow."

Jungle Patrol

A SETTING AND A MOOD

Thus ended the bloodiest encounter of the Spanish-American War.

There are other indications that the conquest of the Philippines was not taken seriously by these casual soldiers of '98. The qualifications for officership must certainly have included a sense of humor. There is a story told of one of the highest ranking colonels, a man afterwards active in the negotiations with the Moros in Mindanao. A very handsome officer he was, with great flowing white whiskers. Army paper work was a bore to this Colonel; his habit was to sign papers as submitted, without examination.

His Sergeant-Major, a youngster of twenty-two, found himself with a lack of funds and decided to capitalize upon this peculiarity of his Colonel. He repaired to a group of his fellows and propounded a most extraordinary wager. For the sum of ten dollars he offered to produce a military document of unusual significance.

His wager was accepted.

It was some days later that the Sergeant-Major approached the Colonel with a great sheaf of orders for signature. The Colonel fiddled irritably with his whiskers as the young non-commissioned officer began an explanation of each order. "This one, Colonel, an order for the purchase of forage for the horses. This one a . . ." The Colonel raised his hand impatiently, squishing with his pen and reading nothing offered for his signature.

Back to the doubting enlisted men raced the Sergeant-Major. He exhibited an official army order, properly signed by the Colonel. "Do you accept this as the Old Man's signature?" he asked. A nod of assent. Whereupon the Sergeant-Major cleared his throat and read joyfully:

Headquarters, —Regiment,
Manila, Philippine Islands,
December 16, 1898.

I, Colonel —— ——, commanding the —th Regiment, United States Infantry, do hereby sentence myself to be shot at sunrise tomorrow. I further direct that the firing squad shall be in charge of Sergeant-Major —— and that the body shall be drawn and quartered and displayed as an example of tropic senility and hardening of the arteries.

—— —— Colonel, Commanding.

Nor was this post-war levity confined to the troopers in Manila, for we see Congressman Bede rising to the floor of the legislative halls of the nation with the ribald suggestion that "America relieve herself of the Philippine problem and at the same time preserve the protective principle by exchanging the Islands for Ireland and then be able to raise her own policemen."

But the levity of '98 was soon to give way to the grim realities of '99. The war with Spain at an end, the head bookkeeper in Washington, D. C., brought the accounts into balance. He determined the interesting fact that the sinking of a few obsolete Spanish ships and the investment of Manila had been accomplished with the trifling expenditure of $300,000,000.

This was a figure to be pondered. And it *was* pondered, for the administration went into session and decided to complete the Philippine deal with the purchase of the Islands from Spain for an additional $20,000,000. Nobody could say America walked in and helped herself; we paid cash. Altruism then began a slow fade into the background and Senator Henry Cabot Lodge stripped the veil from all pretense very honestly when he said, "We make no hypocritical pretense of being interested in the Philippines *solely* on account of others. We believe in trade expansion."

And so, at the will of one man, the President of the United States, the jovial days departed and the frightening days of conquest and consolidation began. McKinley came into the open then, and for the first time the Filipinos realized that they were in process of confiscation. The President made his aims quite clear: "While we are conducting this war and until its conclusion, we will keep all we get; when the war is over, we must keep what we *want*."

With the purchase of the Islands from Spain, the United States contracted for the well-being of some eighty tribes of Malays. They contracted also for the pacification of the Philippines. They were to find the Filipinos singularly irresponsive to the suggestion that they accept annexation at the hands of a Western democracy. The United States was to experience a period of chill and ferocious guerrilla warfare that was to reach across decades of time, and they were to learn that the regular troopers of the Army of the United States were not adaptable to the dripping jungle that was the terrain.

What of these Filipinos?

They are a polyglot people of strange and conflicting Malay, Indonesian, and Indo-Australian tribes, intersprinkled with a dash of the blood of a dozen races of the East. In the north is to be found a linking affiliation with the Chinese, and everywhere is to be seen traces of the curious little Negrito who, apparently, was the original settler of the Islands.

Jungle Patrol

A SETTING AND A MOOD

The Filipino, withal, is a sturdy jungleman, easily aroused to fanaticism and possessing a vast pride of race. He is a doughty fighting man, as is his cousin the Moro, of the south. Mix together a Protestant, Catholic, Mohammedan, and idol-worshipping population (if this be possible); concede a hot-blooded touch of paprika to the veins, and a great facility with edged weapons; to this add a great intertribal distrust, a confused babel of eighty-seven dialects, and a considerable oratorical ability in any one of these dialects: the product is a Filipino.

Enrich this mixture with a passionate love of the ideal of liberty and an extreme readiness to die for that ideal, and the nature of the native resistance to America is clarified.

The resistance to the United States in 1899 should not be belittled by the term "Philippine Insurrection." It was more than an insurrection—it was a legitimate war of protest, waged under capable and idealistic Filipino leaders. Mabini summed up the Filipino aims of 1899. He was a great Filipino statesman, and his words reflect his greatness:

> "The Filipinos realize that they can expect no victory over the American forces; they are fighting to show the American people that they are sufficiently intelligent to know their rights. . . the Filipinos maintain their fight against American troops, not from any special hatred, but in order to show the American people that they are far from indifferent to their political situation. . ."

We have another picture from the past to paint the purity of this Filipino resistance in that early period before they turned to banditry and arson and murderous guerrilla warfare. This one from the pen of Richard Henry Little, an American war correspondent who recorded for the *Chicago Tribune* the death of the Filipino general, Gregorio del Pilar:

Jungle Patrol

A SETTING AND A MOOD

It was a great fight that was fought up there on the trail of lonely Tilad Pass on that Saturday morning of December second. It brought glory to Major Marsh's battalion of the 33rd Volunteer Infantry who were the victors. It brought no discredit to the little band of sixty Filipinos who fought and died there. Sixty was the number that at Aguinaldo's orders, had come down into the pass that morning to resist the onward march of the Americans. Seven were all that went back over the pass that night to tell Aguinaldo that they had tried and failed. Fifty-three of them were either killed or wounded. And among them, the last to retreat, we found the body of young General Gregorio del Pilar.

We had seen him cheering his men in the fight. One of our companies crouched up close under the side of the cliff where he had built his first entrenchment, heard his voice continually during the fight urging his men to greater effort, scolding them, praising them, cursing, appealing to their love of their native land and the next instant threatening to kill them himself if they did not stand firm. Driven from the first entrenchment, he fell back slowly to the second in full sight of our sharpshooters and under a heavy fire. Not until every man around him in the second entrenchment was down did he turn his white horse and ride slowly up the winding trail. Then we who were below saw an American squirm his way out to the top of a high flat rock and take deliberate aim at the figure on the white horse. We held our breath, not knowing whether to pray that the sharpshooter would shoot straight or miss. Then came the spiteful crack of the Krag rifle and the man on horseback rolled to the ground. When the troops charging up the mountainside reached him, the boy General of the Filipinos was dead.

So this was the end of Gregorio del Pilar. Only twenty-two years old, he had managed to make himself a leader of men while he was hardly more than a boy and at last he laid down his life for his convictions. Major Marsh had the diary. In it he had written under date of December second, the day he was killed: "The General has given me the pick of all of the men who could be spared and ordered me to defend the pass. I realize what a terrible task is given me. And yet I realize that this is the most glorious moment of my life. What I do is done for my beloved country. No sacrifice can be too great."

A private sitting by the fire was exhibiting a handerchief. "It's old Pilar's. It's got Dolores Hoses on the corner. I guess that was his girl. Well, it's all over with Gregorio."

"Anyhow," said Private Sullivan, "I got his pants. He won't need them any more."

The man who had the General's shoes strode proudly past, refusing with scorn an offer of a Mexican dollar and a pair of shoes taken from a private insurgent soldier. A soldier sitting on a rock was examining a golden locket containing a curl of a woman's hair. "Got the locket off his neck," said the soldier.

As the main column started on its march for the summit of the mountain, a turn in the trail brought us again in sight of the insurgent General far down below us. There had been no time to bury him. Not even a blanket or poncho had been thrown over him. A crow sat on the dead man's feet. Another perched on his head. The fog settled down upon us; we could see the body no longer.

> "We carved not a line and we raised not a stone,
> But we left him alone with his glory."

And when Private Sullivan went by in his trousers, and Snider with his shoes, and the other men who had the cuff-buttons, and the Sergeant who had the spur and the Lieutenant who had the other spur, and the man who had the shoulder straps, and the other who had the handkerchief, it suddenly occurred to me that his glory was about all we had left him.

With the death of Pilar and the surrender of Aguinaldo, a new kind of warfare came to the Philippines; an unpleasant warfare, without idealism or any purity of motive. As the bona fide insurrection muted away to an undertone, the United States faced the subjugation of brigand and pirate and the conquest of an oozing jungle. A frightful bush made up the terrain that was to be the battle scene.

The Philippines are an archipelago of many names and of piled-up history. We scan it now, on the maps, as a chain of tropical islands that were torn loose, long in the past, from the mothering east coast of Asia. They are islands that form a coral barrier to the progress of the big blue rollers of the Pacific, and they mark the line of the dread malaria zone that separates the East Indian Archipelago from the pleasant isles of mid-Pacific where the Polynesians dwell. 3141 islands, they say, of which only 1473 have been awarded the dignity of names.

On clear days, when the sunlight traces and foreshadows every coral reef and sand-lipped lagoon, one can stand on the southernmost edge of the Japanese island of Taiwan, at South Cape, and look across to the horizon shadow of the most northerly of the Philippine Archipelago. Those islands are the Bashee Group, lying twenty-two degrees north of the equator in the very fringe of the Empire of the Rising Sun.

From that northern reach, the Philippines stretch southward for a thousand lazy, sun-drenched miles, until they end, reluctantly, in the very Bay of Darvel. Little Sitanki, the last of the lengthy chain, is separated by a fifteen-mile strait from the frowning, jungle-wrapped frontier of British North Borneo.

The Philippines have had many names and many masters. Or possibly we should say, have mastered many. America came late to these islands and lingered briefly.

Jungle Patrol

A SETTING AND A MOOD

Others had found them, fought over them, named them—and found death in them. No one can say where the strange, unwritten history of the Philippines began. The earliest peoples have left no kindly records for the perusal of the scholar or the historically curious. They wrote their records in red, with bolo blades and wavy-edged krises, on a shining white beach. But the tides came, and the tropic rains, and washed the records away.

Some few speculative facts remain, to link the Islands with a past that seems incredibly ancient. We know that the men who wrote Sanskrit came to this archipelago—their characters survive as the possessions of exotic tribesmen of the islands of Palawan and Mindoro, inscriptions incised with fire on bamboo.

The islands have been the goal of many and the reward of few. The grave Ming emperors called them the Islands of the Luzones, and sent to them trading junks with brassware and porcelain and silken cloth and little copper bells. Returning to China, these junks carried pearls and precious wood. Ptolemy, the great Egyptian geographer, gave them a resonant name. The Maniolas, he called them, and the Phœnician traders cruised their coastline in quest for gold. In their turn, the wandering Portuguese, greatest of all navigators, charted them as the Islands of the West.

Then came Magellan, their official "discoverer," to give them their first historical title in 1521—The Archipelago of Saint Lazarus. Saint Lazarus Isles,—a fine, round name that perished when Magellan fell before poisoned arrows on Mactan Island. The disorganized tribesmen we call Filipinos might have been "Lazarites" had Magellan not died too soon.

But in 1543, came Ruy Lopez de Villalobus to give them yet another name, and to strangle at birth what might have been a great battle cry. Las Felipinas he called them, in honor of Don Felipe, Crown Prince of Spain.

So the name has remained, with charming variations. Thomas Candish came in 1588, to add a note of sonorous intonation in the grand manner of the Old English tongue. He dignifies the Islands with phrases of great beauty:

> From the Cape of California, being the uttermost part of Nueva Espagna, I navigated to the Islands of the Philippinas, hard upon the coast of China; of which country have I brought such intelligence as hath not bene heard of in these parts. The statelinesse and riches of which countery I feare to make report of, least I should not be credited; for if I had not knowen sufficiently the incomparable wealth of the countery, I should have bene as incredulous thereof, as others will be who have not had the like experience.

It was thus that the Philippines emerged from anonymity to be named, mapped, and ripened for conquest. It was from that pleasant historical siesta that they awakened to find themselves the center of grave Senatorial debates and the object of outraged yelpings of indignant sugar-beet planters.

With the outbursts of sugar lobbies in Washington, D. C., and the shrill cries for *independencia,* this volume is not greatly concerned. This is a story about men. When the political confiscatory measures were completed and the honest insurrection of the Filipinos quieted, these men came to make safe the country for American and Filipino alike. Theirs was the job of mending a country that had been wracked by war and was now subject to the raids of pirates and plunderers and murderers. They were to be the supervisors of the public peace and the strong right arm of the new American civil government.

In 1901, the Philippine Archipelago lay sullen under

the unwanted rule of a Western civilization. The glamorous days were no more; with their passing, the sympathetic phases of the Filipino resistance were passing too. Young Gregorio del Pilar, the magnificent, was gone now. Aguinaldo, the flame of the rebellion, had taken the oath of allegiance to the United States. With their passing, something fine went out of the Filipino resistance and the heart of the real insurrection was stilled.

Pilar and Aguinaldo were replaced by bandits and religious fanatics.

As the contest for sovereignty wore along through the dreary months, a new breed of men came to the scene. We begin to see them spreading across the face of the archipelago, a new kind of fighting force with a strange and fantastic uniform. They were greater, this new breed, than the political system that made necessary their existence; they were far greater than the confiscatory era to which they belonged.

They were policemen of the jungle.

They were bush-bred men with a facility for surviving ambush and death traps of the jungle. Many of them were carry-overs from the Indian wars in the United States that had opened the last American frontier. They understood the ways of wild people. They knew how to fight—and they relished odds. They risked their lives with an elaborate, casual carelessness. They were large men, with drooping mustaches; they spoke slowly and they were adept with rifle and pistol; they wore uniforms of linen with red epaulets: they were officers of the Philippine Constabulary.

Chapter ii

"*REMEMBER BALANGIGA*"

> "The object of patrols is to capture or destroy ladrones, to guard and police the country and the routes of travel . . ."
>
> —*Constabulary Manual*

IT is the afternoon of February 24, 1902. Inspector Henry Knauber of the Philippine Constabulary is en route with two enlisted men from the town of Indan to the city of Magallanes, which is in the Province of Cavite, Philippine Islands. The Inspector is a youngster of twenty-six, a German, born in Berlin but now a citizen of the United States. With the outbreak of the Spanish American War he had enlisted as a private in the 2nd Artillery, and at the suspension of those casual hostilities he had taken a discharge in Manila, as a Sergeant of the old 32nd Infantry, United States Volunteers.

During his period of civil residence in Manila a new quasi-military organization has come into being, and Knauber has been one of the first to accept the rank of Second-Class Inspector of the newborn Philippine Constabulary.

He is riding along, this afternoon in 1902, with two soldiers at his back, en route to a new station. The Philippine jungle of this day is alive with malcontent *insurrectos,* religious fanatics, and unadulterated bandits. The traveler, in passing from village to village, keeps his weapons at the ready. It is a land where men live or die according to the proficiency of their arms. The way of Knauber leads along a twisting trail that is bounded by tall *cogon* grass and shaded by the lofty tops of the buttressed forest trees.

Knauber is thinking, possibly, that the place is admirably suited for ambush. He rides cautiously, his outmoded Remington shotgun at the alert.

A keen-eyed little brown private spurs his small native pony forward to speak to his officer. "Señor, there is a glitter in the grass."

"*Ach,*" said Knauber, "ve haf trouble, maybe?"

The hoofs of the ponies skid in the dusty trail as the detail reins to a halt. The jungle is very silent. Too silent. For a moment they peer away across the billowing grass and into the folds of shadow that creep along the jungle edge. Then they ride on again—more slowly.

The whine of a Mauser bullet salutes them at a bend of the trail. Smoke drifts across the tasseled tops of the *cogon* as the shotgun in Knauber's hands blasts at a whirling shadow that rises from the grass. A wild shriek, bloodcurdling and shrill, sounds in the depths of the grass as eight insurgent soldiers of the disorganized Filipino army rush the Constabulary detail.

The horse Knauber is riding collapses with a scream. The Inspector disengages his feet from the stirrups as he falls, and dives for the safety of the bush. His two men abandon their horses and take cover with him.

The rush ends as suddenly as it began. Men again dissolve into bush. Knauber and his soldiers crouch there in the grass in a silence that presses against their imaginations. The *cogon* is shoulder-high about them, and they experience an elemental moment as they visualize grim little brown stalkers squirming in their direction. They wait. There is nothing else to do.

Then the Mauser rifles begin to speak again, the smokeless powder shielding the position of the unseen riflemen. Bullets begin to clip the tufts as the insurgents seek their quarry. The police detail takes the rifle fire in silence, scanning the green aisles with pounding hearts as they prepare for the inevitable rush that will bring men hand to hand.

It comes. With a screech, the attackers hurl themselves through the matted grass roots; keen blades are swishing as they chant the dreadful, monotonous cry of the Filipino fighting man. The three Constabulary soldiers rise to meet the rush of eight bolomen who are commanded by one "Capitan" Julian Ramos. There is a blur of sound and movement as men face the greatest adventure of all.

Teeth and shotgun butt and bolo edge and bare hands come into play that day, and when fifteen frenzied minutes have passed the jungle becomes silent again but for the clatter of the indignant monkeys and the protests of the noisy parrots.

For thirty-five years they have remembered what Knauber said that day as he mopped his brow and surveyed the job with a pardonable pride. *"Gott!"* said Knauber, beaming at a private. "It was most uncomfortable, yes?" And then he rode on again to Magallanes with his two men!

Incredibly, he had emerged from that jungle dogfight alive—his two men at his back. At Magallanes, he made a modest report:

> "Near the *barrio* of Caititinga my detachment was rushed by eight men under the leadership of Captain Julian Ramos. We killed eight, captured two Mauser rifles, one Remington rifle, one revolver, bolos and one trumpet. Constabulary casualties, one horse killed."

This engagement was representative of the thousands of hand-to-hand combats that resulted in the pacification of the Philippines. It was unusual only in that it resulted in the first awards of the Constabulary Medal of Valor. For this action, Second-Class Inspector Henry Knauber and Privates

Jungle Patrol

"REMEMBER BALANGIGA"

Manuel Gonzales and Luis Perez were given the highest award the Philippine Insular government could offer. They thus became the first wearers of that ribbon whose color is crimson, with thirteen white stars woven into the field. By General Order number 8, dated February 27, 1902, they were so honored.

The Philippine Constabulary, at the time, was six months old. A great part of this half-year of its existence had been taken up with organization details and with recruiting and examinations; but in that short interval, six men of the new Insular Police force had found opportunity to die for the peace of the Philippine Islands.

The history of the American occupation and conquest of the Philippines is, in large measure, the history of the Philippine Constabulary; for it was this force of native infantry that applied the finishing touches of civilization to a jungle land that had known no law. No word picture could approximate the chill drama that was the daily portion of these jungle police of Uncle Sam as they went about the business of applying that civilization.

The Constabulary fought sometimes in collaboration with the troops of the United States Army; again, it fought as isolated patrols in the bush, cut off from all human contact and from all source of supply. And sometimes, it is sorrowful to relate, it fought with neither the support nor the cooperation of the armed force of the United States regular army.

At organization, the Constabulary was an unwanted, orphan child. Its birth was a military necessity, and its actual formation was undertaken only after prolonged and acrimonious debate. In the early days of its existence the loyalty of the Constabulary was questioned, and there was great protest in American military circles against this arming of natives.

It was argued, and rightly, that there were already too many rifles in the jungle!

It was said that the Constabulary would be a focal point for organized resistance against the Government; it was said that it would be impossible to discipline natives into a smooth, efficient corps; and it was said that the Filipinos would desert with their rifles at the first opportunity. It was believed, too, that the force could never be more than an ineffectual policing aid to the regulars. Regardless of these uneasy protests, the Constabulary was a political necessity and it was organized and carried through to existence in the face of all dissension.

It was intended, from the first, that the Constabulary should inherit the dirty work.

There was a very valid reason for the opposition of the regular army to the creation of this force of native infantry. The experiences of the army with Filipinos had not been pleasant ones in 1901. Conversation in the army messes was centered upon men who had been buried to the necks in ant hills. The army did not agree with the perspiring and benevolent Mr. Taft, who spoke so fondly, if vaguely, of the "little brown brothers." The army had a song to disclaim all relationship with the Filipinos:

> "I'm only a common soldier man in the blasted Philippines,
> They say I've got brown brothers here but I don't know what it means,
> I like the word fraternity, but still I draw the line—
> He may be a brother to William Howard Taft, but he ain't no brother of mine."

Jungle Patrol

"REMEMBER BALANGIGA"

The military objection to the arming of native police was well supported by the War Department reports that contained chill tales of ambush and treachery and jungle intrigue.

The army forecasts of disaster were not realized. Not only did the Constabulary prove loyal to the government, but its record of lost arms, desertions, and renegades was to compare favorably with the record of the army. There *were* Constabulary renegades, but many of the prime trouble-makers came, not from the files of the Constabulary, but from the ranks of Scouts, Volunteers, and regular army.

In the field, the Insular Police demonstrated a principle that was old and well understood by the British administrators of Colonial India—that native troops, thoroughly understanding the terrain and the psychology of the population, were more effective than white soldiers in the conduct of jungle warfare. It is a common-sense principle.

It is a matter of record that the morale of the Army of the United States in the Philippines had reached a low ebb in the year 1901. In massed movements against the organized Filipino army it had been most effective. As jungle patrolmen, the effectiveness of the army was not so apparent. In 1901, the official insurrection was finished, but there remained the onerous duty of hunting out phantom bands of armed irreconcilables who had refused to accept American authority. Nor was this a period of brigandage or banditry. It was an era of unofficial, insurgent warfare, with uniformed detachments of Filipino troops engaged in guerrilla operations.

These ghost detachments vanished too easily into the screening bush when the regulars took the field.

The ponderous troop movements of the army resulted in

severe setbacks in 1901. The regulars were unable to adapt themselves to service in small detachments and to the carrying of their homes on their backs on extended jungle patrol. They were not too successful in their efforts to live on the bounty of the bush. The army needs commissaries and ammunition trains and hospital corps; things that were lacking in the Philippine bush.

The operations had been disastrous, not only in cost of lives but in damage to military prestige. The massacre of an army patrol would encourage to life a dozen new revolutionary movements. Even in those early days, the Filipinos were formulating their policy of "Better Hell under native rule than Heaven under the *Americanos*." The hatred for America was expressed then with bolo blades rather than with independence missions. That was the difference.

The army had incidents of jungle warfare to remember. . . .

There was the story of an American patrol party in the bush of Luzon. They had been somewhat inexperienced in the not gentle art of *bosque* warfare, and had speedily lost all sense of direction in the clinging bush.

A scouting party of three men was detached and sent ahead to find the trails, if any. They did not return, and the rest of the force straggled on to emerge, after a severe interval, to safety. But they did not forget those three scouts, and patrols went out again to search that tangled bush that had engulfed them.

Days later, they came to an abandoned campfire that had been an outpost of the Filipinos. There they found their lost men. The terrified old woman they rounded up said that the Filipino officers had sat on the porch to laugh at the frenzied antics of three men, buried to the waist, who had sought to fight off the millions of swarming ants.

It was a bad story; it left a taste in the mouths of volunteer and regular alike. And it broke the rules.

Another tale concerned an early army experience in the Mohammedan country to the south. There the army had been building a road through the jungle, and a young Sergeant had taken his duties too seriously as a foreman of the road gang. One day he indicated a shovel to a proud *Datu* of the Mohammedans who was standing there, erect and aloof. The Mohammedan chief ignored the suggestion that he soil his hands with manual labor. The Sergeant lashed out with his boot.

A flicker appeared in that impassive Mohammedan's face. His hand tightened on his kris, and for an instant fire lighted the fierce eyes. An ancient American packer, jungle-wise and illiterate sauntered by to witness the scene. His great beard was stained with ill-directed tobacco juice and he was an object for laughter to the spick-and-span regulars. But he offered a note of excellent advice to the too earnest Sergeant. "Kill 'em after you kick 'em," he advised casually, "or they'll git you, Son, sure as Ol' Billy Hell."

The Sergeant grinned as the old-timer spat against the wind and sauntered on. The next morning, the Sergeant was still grinning as he lay beside the road. His head had been carelessly kicked away from his torso.

Sometimes it seemed that even the jungle itself conspired against these raw young troopers of the regular army who knew so little of the bush land. The messes had a story of a recruit. . . .

The boy had been nineteen. Life had been an adventure, to be lived every day, while one was yet nineteen. His father's grocery store in Tennessee had been a stodgy place; pudgy drummers leaning over a cigar counter to tell risqué stories about farmer's daughters.

Jungle Patrol

"REMEMBER BALANGIGA"

So the boy had gotten away from it all; had come to the Philippines . . . army man now, at nineteen.

They put him on sentry post in the graveyard watch—four o'clock until dawn. He must have felt the romance of his job as he gripped his Krag and stepped out along the shadow-darkened path beneath the curling tree-limbs.

They came for the boy at dawn—the guard relief—and he was gone. His hat and rifle were there on the ground near his deserted post, and the soldiers walked along that line that separated them from jungle until they found the sentryman who had been nineteen.

Python. . . .

Turn back, if you will, to the morning of September 28, 1901. Company C of the 9th Regular Infantry is on station at Balangiga on the south coast of Samar Island. Their barracks are fringed with a wall of frowning jungle, and that jungle is alight with the silent flames of rebellion. These boys, from North Carolina or Montana or Maine, know too little about that reaching *bosque;* and far too little about the strange, wild people who inhabit it.

Looking back to that station, from the advantageous hindsight of thirty-five years, we know, now, that they were careless. Company C of the 9th Infantry was newly arrived from China . . . and China is so different a country from the jungle of the Philippines. These cocky, confident regular troopers underestimated the resourceful, savage killers who frequented the bush of Samar Island.

In the barracks, a song is in process; one of the many songs that make up the undertones for all of the campaigns of America:

Jungle Patrol

"REMEMBER BALANGIGA"

"Underneath the starry flag,
Civilize 'em with a Krag,
And return us to our own beloved homes."

Soon they are to die—because of these same Krag rifles.

In the supply room a Sergeant is making an early morning check of equipment. It is 6:45 the sun is barely over the rim of the jungle. Mess call blows and the troopers pour into the company street. They are superbly confident in the possession of their repeating Krags—even in the face of that sunlit but somber jungle. They line up with their mess kits, buoyed by the safety of numbers.

There comes a wild clamor of the church bells in the little town; an unwarranted and unauthorized clamor and it causes the troopers to spring for their arms. A bugle blares —shrill against the clamor of the bells. The soldiers mill about in the company street, the voices of the officers hoarse above the din. Balangiga has changed suddenly from a peaceful little tropic *barrio* to a place of grim and chill menace. But even then the milling troopers feel that there is a mistake—the "niggers" would never rush a whole company of regular infantry in barracks!

Then comes the thud of steel blade on living flesh as the sentries gurgle their lives away. The Krags have no time to rattle before Company C is rushed, front and rear, by 450 natives who burst from the deadly concealing bush.

The regulars had not a chance for life. Careless in their fancied security, many were unable to reach their rifles. The men fought almost barehanded; one of the soldiers killed several of the attackers with a baseball bat before he was overwhelmed. The cook falls across his fire—bleeding his life away over the popping coals.

It is all over too soon. Captain Thomas W. Connell goes

down beneath the swishing blades, his officers about him. Sergeants take command, to fall in turn. . . .

The next day Captain Bookmiller enters the silent town. The hacked limbs and shattered torsos of men greet his horrified eyes. Grimacing heads stare from the corridors—heads without bodies. The attacking force is gone—with them, 100 Krag rifles and 25,000 rounds of ammunition. Bookmiller buries the dead: forty-eight mutilated bodies. He burns the town and retires.

The next day there stagger into Basey twenty-four men, eleven of them wounded. They are all that remain of Company C of the 9th United States Infantry.

For many years, the army will "remember Balangiga."

Less than a month later, on October 16, Company E of the same regiment comes under attack on the Gandara River in Samar. Here, at Camp Denver, the troops are also aligned at breakfast when they are assaulted by 100 armed fanatics. With white baggy trousers blowing in the wind, and bolo blades glistening under applications of coconut oil, the attackers all but overwhelm the American infantrymen. Only the rapid magazine fire of the Krags saves the company from annihilation. Here the attacking force numbers but *one* hundred, to make the odds more nearly equal. Twelve minutes of battle and the natives withdraw, having accomplished the killing of eleven American troopers and the wounding of six.

We have an almost forgotten historical incident to remind us of the slaughter at Balangiga, an incident that has survived only in the memories of field officers of the early 1900s. It is to be found in no official order of the army of the United States but it concerns an official reply of Colonel Hughes of the 9th Infantry.

To this Colonel, General Adna Chaffee wrote, following the massacre at Balangiga. In substance he said, "It comes to my attention that Company C of the 9th Infantry was very poorly equipped in the soldierly essentials of discipline, training, organization, and morale. Your statement is awaited."

To which Colonel Hughes of the 9th Infantry replied, "It may be true, General Chaffee, that Company C of the 9th Infantry, was lax in discipline, training, and morale. I would not be qualified to comment upon the subject as the Company has been under my command for but two weeks. During its previous China service, it was under the direct command of General Adna R. Chaffee, Commanding the Philippine Division!"

To which, it was stated, General Chaffee had no ready reply.

In Manila this General Adna R. Chaffee, commanding the Philippine Division, prepares a reluctant report. Before him is the miserable news of the slaughter at Balangiga. His orders are conveyed to all commands:

> "We have lost one hundred rifles and 25,000 rounds of ammunition at Balangiga. You must get them back. You can have $5,000 gold. Capture arms if you can, buy them if you must; whatever course you adopt, *get them back*."

And then the General turns to his annual report to Washington, D. C. Quite unsteadily he writes, "I fear that our soldiers, transplanted to a strange sphere of action, do not fully appreciate or realize the difference in their surroundings."

Colonel Charles R. Greenleaf, Assistant Surgeon-General of the Army, adds a line to the report that is for the perusal of the President of the United States. "The most energetic and stalwart American," he writes, "after a year of service here loses energy, strength, and ambition. He performs what duty his work demands more or less half-heartedly."

Even as these reports were being penned, the Constabulary was in organization to fill a pressing need; moving up, as it were, to take over the dirty work of the jungle patrols.

There is a permanence about the jungle. . . .

If it is to be conquered, it must be by men who know its recesses and all of its moods; by men who are as keen as Apaches in reading the position of a careless vine in the trail that marks a spear-trap; by soldiers who know its edible roots and which of its fruits are poisonous. To force the back country with any slight measure of safety requires a knowledge of the native languages and, most of all, the acquisition of the native psychology. To survive, one has to *think* in the manner of a hillman.

The army could not do these things. They relied upon man-power and superior armament to carry them through. Sometimes it *did* carry them through—too often it failed. Never could their arms or equipment compete with native jungle cunning in that tangled *bosque.* It was a job for men who specialized in jungle.

Then, too, the army comes and goes: a hitch, as they call it—two years, possibly three years; then they are gone, to be replaced by new troops to whom all is unfamiliar.

To the army soldier, the job is a hitch; the Constabulary patrolman measured his term of service in decades. He conquered the jungle as much by familiarity as by force of arms. Some of the best of the Insular Police grew old in the shadow of the equator.

The Philippine Constabulary took from the army the tiresome and thankless detail of policing the Islands, and they pacified the jungle by armed infiltration. In so doing, they accomplished that desirable result for which the army was hopelessly unsuited and for which the Constabulary was born.

The Insular Police fought under conditions, and against odds, that make even the efforts of the Texas Rangers seem ordinary police work. Nor was the Constabulary, at any time in its existence, a strong force in numbers, equipment, or armament. Their opponents were for the most part armed fanatics who fought, not to preserve life or an ideal, but for the sheer love of dying. By no reach of the imagination could the casual road agents of the West or the Mexican cattle thieves of the border be compared to the fierce *pulajans* who screeched from the Samar jungle, on a mission of torture and raid. The outlaws of the West did not ambush; they fought with six-guns, in front of a saloon bar, and the best man won. And usually they fought an individual combat, one man against one man.

The *pulajans* will be discussed fully in a later chapter. They were hillmen of the Philippine inner country. They wore shirts of red, marked with white crosses, and they were religious maniacs with the ferocity of a panther and the cruel cunning of a wild boar. They retained a numerical superiority over the Constabulary patrols they engaged. They struck without warning, and they maintained strongholds deep in the jungled mountains. The trails to their citadels were trapped with poisoned spears, poised for release at the jungle edge. Every innocent vine across the path might be a trigger of death. Every carefully placed sliver of bamboo was a threat of septic poisoning.

Every refinement of ferocity was embodied in the tactics of the *pulajans*. One of their standard methods of attack was for single individuals to lie in wait in the long grass beside the jungle trails. As a patrol came, the *pulajan* would lie silently until the last man had passed. Then with the silent speed of a leopard, the *pulajan* would rise, a dagger in each hand, leap into the trail behind the soldier, and embrace him around the middle. Sinking the daggers into the victim's stomach, the *pulajan* would rip backward with each hand, disemboweling the soldier before the attack had registered in the minds of the patrol.

The subjugation of the *pulajans* is one of the most grim notes in the records of American colonial history. It is a record of cat-footed night attacks, with soldiers on terrible sentry duty, bleeding lives away through throats cut ear to ear. It is a story of waving fields of *cogon* grass, where death waited patiently for a patrol. It is an epic of guerrilla warfare with no quarter for the vanquished and small hope for the wounded who lay in that festering bush.

The little brown soldiers who made up the enlisted personnel of the Constabulary, took their oaths of allegiance to Uncle Sam with great seriousness. Private of Constabulary corps, or *pulajan* of the mountain tribes, there can be no question of the fitness of the Malay for martial work. In this breed is to be included the Moro, who is a Mohammedan of the southern islands. All of the tribes are practically immune to fear, and each is possessed with a resistance to disease, exposure, hardship, and exhaustion that can at times be a marvel to a white man. Under the direction of the capable white officers of the period of conquest, these natives were wildcats in khaki, and they pursued their erring country men with the zeal of a hunting cheetah.

When the jungle closed behind a Constabulary patrol, pushing a point deep into the country of *pulajan* or Moro, issues became strictly man-to-man. There was no supporting force in reserve, no commissary to feed starving soldiers, and no hospital train to gather up the wrecked from war.

When the Constabulary came to preserve the law of the white man in Mindanao and Sulu, it was in a region that had obeyed only the Koran and the whims of man. The record of the dark and ghastly years of the Mohammedan patrol duty was the epilogue the Constabulary wrote to the *pulajan* wars of Leyte and Samar.

And in the Moro campaigns too, the Constabulary came through with honor.

With reams of paper devoted to the activities of the French Foreign Legion, the Texas Rangers, the International Column of Spain, and the Royal Northwest Mounted Police, it is amazing that the jungle purge of the Philippine Constabulary has missed the attention of the writers of the battle memoirs of our nation. It *was* the Foreign Legion of America! Officered by Americans, Englishmen, Germans, Spaniards, and a few Filipinos, the red epaulets of the corps took on an international significance.

The Foreign Legion of America they truly were, and at the same time the agent of the American civil government in the Philippine Islands.

The Constabulary was the victim of a vicious political condition. It had been the policy of the American colonial administration of those days to minimize the disturbances in the Philippines and to withhold from circulation in the United States any knowledge of the considerable native rebellion against American authority.

Official communications to the American people were couched, in 1900, in terms of depreciation of the insurgent activity. "The Philippines are peaceful and American intervention is everywhere welcomed." That had been the tenor of the notes prepared for the people of the United States.

With this background of political misinformation, it became impossible to order out the regular troops of the army to put down active insurrection and murder and arson. The very presence of the regulars in the field would have been tacit admission to the American public that all was not well in the Philippines. And so to bring a restless public into line, the army was ordered to barracks too soon. From those barracks, they watched the smoke of burned villages and saw the flight of a demoralized civil population before the raids of *pulajans* and Moros.

It was to fill this breach that the Insular Police came originally into being, the object being to preserve public order without the presence of official troops in the field. The Constabulary was intended to be, not a military body, but a policing unit: their duties—to put down armed disturbances that should have been put down by the army. The country was not pacified, and there was no valid justification for civil government in 1901; nor was there the slightest chance of this government succeeding except under the pressure of some armed force. The Constabulary was thus a sub rosa body of unofficial American soldiery, created to quell, with discretion and without publicity, a very serious public disorder. As the Filipino resistance was denied official confirmation by the Philippine civil government, the Constabulary was undertaking a grim, unpublicized warfare.

This police force came into being under the name "Insular Constabulary," and the corps was greeted with amusement and a great disdain by the professional soldiers of the regular army. And possibly with reason, for they must have been a nondescript group in those days. Their uniforms were ragged, their training was nil, and their armament was obsolete. It was with exceedingly poor equipment and with small organization that the force was born to its work.

The Constabulary was handicapped, too, by its very corps name. The "I. C." of its collar ornaments was a source of great laughter; for the term "I.C.," in army parlance, means "Inspected and Condemned." So it was as the "condemned corps, without inspection" that the Constabulary took the field. The name remained until the administration mercifully changed their rating to Philippine Constabulary, and replaced the offending collar ornaments with the initials "P.C."

They were organized without regiments, it being the intention to use them exclusively in small units of companies, platoons, and squads. Such was the organization that they retained to the end. Always they were split into small, mobile detachments, capable of living off the country and able to persist on detached duty without benefit of supporting columns or ammunition trains.

The Constabulary soldier carried his home on his back. The scene of his warfare was a dripping bush,—a jungle that fed him, starved him, concealed him, ambushed him—and quite often buried him.

In its entire history, there is no record of mass movement of troops. Occasionally companies combined when the odds became too great, and sometimes there was collaboration in the mechanics of forming the great circular cordons. But mostly their work was company against Moro *cotta;* platoon against *pulajan;* squad against *juramentado* Moros.

The manner of their fighting has gone forever now, with the birth of the machine gun, the grenade, and the poison-gas shell. Fighting policemen they were, who developed into the most efficient of jungle warriors. Possibly there are few bodies of jungle soldiers who have attained their state of efficiency in the whole long history of colonial development.

The investigator who seeks their written reports reads them casually at first; then blinks his eyes to read again—carefully. Incredibles in print, might be his first verdict as he surveys their combat years. Their story is an almost unbelievable one of fortitude in the face of danger and warfare against terrific odds. Their battles were fought with every physical and topographical advantage favoring their enemy. Their patrols operated miles from any base, in the center of hostile country. In garrison, they were under constant attack and they suffered from a shortage of munitions and supplies. They were the police of the richest nation in the world, but they were woefully unequipped. The deaths of many gallant young Americans in the Philippine bush is an indictment of the laxity and inefficiency of an inexperienced colonial government.

The Constabulary was restrained in battle. For, in their capacity as a police unit, and not as a military organization, it became their duty to hold their fire and to maintain their position by diplomacy if possible. The first shot went to the adversary.

When one considers the terrain over which the Constabulary operated, the martial calibre of the men who opposed them, and the magnitude of their jungle beat, the disciplinary campaigns of the corps seems almost miraculous. With regularity, young Constabulary officers penetrated that wall of bush with a pitifully few riflemen at their backs, to reach deep into savage country and to engage vastly superior numbers. Sometimes they were cut to pieces; sometimes they came out again, with a double victory over mankind and the jungle. The marvel is that one of them lived to see the coast line again.

Jungle Patrol

"REMEMBER BALANGIGA"

In reading of the campaigns of men like Crockett, Furlong, Tiffany, and others of the personnel who are told of in this volume, one is led to the belief that the Philippine jungle produced the last of the true fighting men who scorned all odds. Certain it is that these men were of a breed that we shall see no more. They belonged to the personal combat era.

Civilization and improved methods of slaughtering men have extinguished the swordsman and the lancer as effectively as it has removed the pikeman and the halberdier and the glittering cavalryman of the heroic days. Almost included in that category is the pistolman, who has been submerged in the long-range tactics of today. Only the rifleman remains to remind us of the time when a man fought as an individual.

And so the battles of the Constabulary are stories of squads and platoons—and often of single men who stood in the glare of the bright face of danger. It is a record of men deep in a jungle, surviving by the accuracy of each man of the detail. In that swirl of stark ferocity, every bush was a threat of death; every clump of grass might conceal a krisman. The music of the spear in flight and the puff of the blowgun mingled with the roar of the shotguns and the clatter of the bolo blades.

The story can never be fully chronicled. It must remain unwritten American history. It was unclean, septic warfare, with no quarter. It was a jungle, alight with the sputter of burning bamboo, with villagers and small children and women roasted alike in the embers of shattered towns. One turns the pages blindly, reading these records of men sent into the jungle to kill and be killed. Of men who were learning the synonyms for the most grim of all words.

Stealthy sorties through the high *cogon* grass; the skewering of men like beetles on the hafts of long-handled spears; the murderous flick of a crimson blade, severing a man from neck to crotch; the scream of the amuck and of the *juramentado* Moro, weaving a twisting trail of death through the Constabulary encampments. Fallen tents a shroud for soldiers decapitated by the swish of *barong* and *campilane;* long jungle nights—terror-haunted—with every rustle in the grass a summons to the God of Battles. Kipling's

> "Things jumping up in the grass,
> To scurry away as you pass . . ."

War in the rice paddies, with men to their waists in the swamp mud; war without a front, and battle without that comfortable solidarity of massed troop movements.

The jungle troopers were a loose organization, with no generals ordering tricky flank movements. The Insular Police had no supporting barrages to protect them from the rush of wiry, maniacal natives. It was individual accuracy in the detail that counted. In the Constabulary, the men who shot truly, lived. The soldier whose aim faltered, went down on the blade of a bolo.

Chapter iii

BIRTH OF THE JUNGLE POLICE

"For the purpose of better maintaining peace, law and order in the various provinces of the Philippines . . ."

—*Constabulary Manual*

THE native resistance to American authority in the Philippines may be readily grouped into three overlapping phases. The first of these was the Philippine Insurrection, so-called, conducted by an official Filipino army which took the field in company, battalion, and regimental formation. The second phase was the period of guerrilla warfare that marked the breaking of the insurgents into disorganized bands of brigands and outlaws. The third and final phase, must, of necessity, be divided into two sections, the first of which was the long series of homicidal operations conducted by the "Popes," "Sons of Jesus," and "Messiahs," who organized the outlaw bands into formidable, well-armed plunderers. The second section was the lengthy campaign in Mindanao and Sulu, where the Moros fortified their *cottas* against American penetration.

These three phases of conflict shaded into one another, and quite often simultaneous operations were in progress against Moros, insurgents, bandits, and the "Popes."

The Philippine Insurrection burst into flame on the night of February 4, 1899, when Private Grayson of the Nebraska Volunteers fired the first shot of the Filipino-American War.

His statement:

"I yelled halt . . . the man moved. I challenged him with another halt. Then he shouted 'Halto' to me. Well, I thought the best thing to do was to shoot him. He dropped.

Then two Filipinos sprang from a gateway about fifteen feet from us. Miller fired and dropped one. I saw another one was left. Well, I think I got my second Filipino that time. We retreated to where six other fellows were. I said, 'Line up, fellows, the "niggers" are all through these yards.' It was some minutes after our second shots before the Filipinos began firing."

The war was on. Started by an irresponsible young sentry, it lengthened into a ghastly hemorrhage of jungle patrol duty that was to last for more than two years.

It was during this two-year period of active warfare that the regular army of the United States reached its point of highest efficiency. The operations were on a larger scale, against massed troops, and the army accomplished the killing of many thousands of Filipinos with the loss of comparatively few men. The official records of the War Department show a loss of men killed in action during the period from February 4, 1899, to July 4, 1901, of but 752. This was in the period immediately *preceding* the formation of the Philippine Constabulary. At this time, the army was at its point of high strength in the Philippines, with a force of 71,528 men under arms. It becomes apparent, then, that our regulars, against an official Filipino army, had been a most efficient body of warriors. The loss of 752 men is very small.

This figure is significant, and will be referred to hereafter in evaluating the part played by the Philippine Constabulary.

The insurrection ended officially in November, 1899. Unofficially and actually, the sputter of insurrection persisted until the middle of the year 1902, at which time it was replaced with a definite era of banditry. But in November, 1899, the American administration believed the pacification to be complete. General Otis stated his convictions in print in emphatic manner. "The country has been pacified," he wrote. "There will be no more real fighting in the Philippines."

But General Otis was looking, not at conditions as they existed, but at the uncertain temper of the American people, who were not sympathetic to the colonial aims of President McKinley. And McKinley had expressed those aims quite definitely. There was no question but that we were headed for the confiscation of the Philippines.

The carefully prepared, misleading statements of General Otis were issued from behind the screen of a severe military censorship that he had set up in Manila. Nor could Otis be blamed for this. He was an army man, and the army acts under the orders of its commanding chief.

American correspondents on the spot were unable to break through this wall of censorship to convey to the people of the United States knowledge of the formidable conditions of unrest that actually prevailed in the Philippine Islands.

But newspaper men will find a way and these correspondents were no exception. Otis stands on his orders, which are to refuse transmittal to all messages "that will be prejudicial to the forthcoming election of McKinley." The correspondents mail their statement secretly to Hong Kong, and from there it is dispatched by cable to the United States, to be published in the *Review of Reviews*.

When General MacArthur succeeded General Otis, he was keenly aware of the dangers of this fragmentary insurrectionist army that was breaking up into guerrilla bands. His writings also indicate that he considered the condition not brigandage, but a continuation of the insurgent resistance. "The disbandment of the Filipino army," he writes, "was not considered in the nature of a calamity by the natives but simply as a transition from one form of action to another; a change which was regarded by many as a positive advantage and was relied upon to accomplish more effectively the end in view."

General Chaffee, who replaced MacArthur, stated that in June, 1901, it was "unsafe to go three miles from Cebu, where two companies of regular infantry were stationed."

So, when the American government offered amnesty to outstanding insurgents in 1900, the offer was very generally ignored, and our government awoke to the unpleasant fact that the insurrection was not yet finished. Filipino secret societies were springing into being—some frankly military, others semi-religious, all intensely anti-*Americanista*. Many of the old *insurrecto* units were still in the field, armed, uniformed, and desperately ready to resist American confiscation. Allied with these were the guerrilla bands who used patriotism as a cloak to justify their lawless acts of pure banditry.

The northern islands of the Philippine group were in a state of ruin, with anthrax destroying the work animals and cholera sweeping the population. It was a time of bitter aftermath of war. The coconut estates were in disrepair, with buildings fallen and rotting away. The noncombatant population was caught in a vise between the terrors of starvation and the marauding bands of guerrillas who had shown no inclination to resume again the arts of peace and agriculture.

Meanwhile, a glance into the bush in 1901 revealed ominous conditions. Isio was there, for one; a wrinkled little Filipino with a delusion of grandeur and a vast following of rabid and frenzied bigots. From his mountain-top citadel he was mentally surveying his Kingdom of Negros; fingering his tinsel crown. . . . Pedro de la Cruz was building up his banditti, who were to lay a trail of red across the archipelago; Rios, the "Pope of Tayabas," was commissioning his "Major-Generals"; Simeon Ola was carrying torch and terror across the vast hemp estates of Albay; Guillermo was unknowingly threading his bloody way to the gallows.

The island of Samar, in 1901, was in process of spawning the most formidable organization of religious desperados that American arms were ever to oppose. Elsewhere in the middle islands, the native secret societies were gathering adherents and flaming with fanaticism.

To the south, in the Moro country, the prestige of America was unfelt. To the natives of Mindanao and Sulu, the United States was not even a name. To the Moros, we were but new white cannon fodder to replace the battered Spaniards. In the mosques, the grave *Imams* were gathering the corded krismen about them, exhorting in the name of Allah. Their voices were droning as they chanted a battle refrain that was old when America was unborn:

> "O ye who believe, when ye meet the marshalled hosts of the infidels, turn not your backs to them;
> Who so shall turn his back to them on that day, unless he turn aside to fight, or to rally some other troop, shall incur wrath from God; Hell shall be his reward and abode and wretched the journey thither."

A battle front was flaming across sixteen degrees of latitude. . . .

But it was at home in the United States that the military authorities were confronted with the greatest problem of all. The people of the United States were singularly apathetic to the colonial aspirations of the administration. And sometimes not so apathetic, as the wave of indignant protests were making it appear. The treaty with Spain had been ratified with but a single vote to break the deadlock, and that vote cast by the Vice President at the direction of the President.

A crucial Presidential election was safely passed.

Then, with more than 70,000 soldiers under arms in the Philippines, the administration undertook to convince the public that the Islands were at peace, and that everywhere the natives welcomed American intervention in their affairs.

But it soon became apparent that a reluctant and stubborn American people would expect more tangible proof that an unpopular insurrection was finished. With 70,000 troopers in the field, the situation was awkward, nor could the presence of this large body of fighting men be satisfactorily explained. It was apparent that something must be done.

Something was done. On July 4, 1901, the military régime in the Islands came to an end, and the government was officially delegated to a Philippine Civil Commission. It sounded much better. It was argued that the establishment of this civil government would make the doubting public look with tolerance upon the colonial project afoot; and to encourage that tolerance, some 25,000 troopers of the regular army were ordered to the homeland. The effect was precisely as McKinley had anticipated. The sight of the returning warriors lulled the suspicion in the public mind and made it easier to forget that there were still some fifty thousand American soldiers in the Philippine bush.

When the Commission, headed by Governor Taft, arrived in Manila, they were informed by the General commanding the Philippine Division that the time was not yet ripe for the setting up of a civil government. The country was in an uproar and had need of the stern policies of a military régime. Governor Taft was inclined to agree with them after a cursory survey of the scene—but orders were orders.

The Commission soon found the statement of the military to be not exaggerated, for they were confronted immediately by three pressing problems. The first of these was how to make the Philippine Islands that place of peace and quiet that had been represented to the American people. The second was how to effect the discharge of a goodly portion of the armed force remaining in the Islands, thereby reducing the cost of the war and stilling the last murmurs of doubt. The third, and most perplexing problem, was how to accomplish the first if the second was to be accomplished.

The fact that all colonial powers maintained a corps of native troops as auxiliary to the regular military establishment was considered at the outset by the Philippine Commission. A tentative step in that direction had been taken with the formation of one company of Macabebe infantry, one company of Tagalog infantry, two companies of Visayan infantry, and a squadron of Filipino cavalry under the command of Lieutenant Batson. The work of these units had been followed with great interest by the members of the Civil Commission. The grand service rendered General Funston by the Macabebes during the campaign that resulted in the capture of Aguinaldo had focused the attention of the administration upon the possibilities of a native force. Too, the army had several thousand Scout soldiers who were an experiment in natives officered by white men.

But the Commission did not want a Federal body of native troops. They wanted a force of unofficial policemen.

What the administration really *did* want was a union of efficiency and economy. The Philippine Insurrection had cost the United States $176,000,000, and it was time that this drain on public resources should cease. It was pointed out by certain thrifty members of the Commission that the average cost of maintaining an American soldier in the Philippines was in excess of $1,000 annually. Some authorities estimated the cost as high as $1,400.

When this figure was multiplied by the 70,000 troopers in the Philippines in 1900 and 1901, the drain on government resources approached the not inconsiderable figure of $75,000,000 annually. Against this sum was balanced an estimated cost of $250 annually for each tentative Constabulary soldier. It was believed that the Constabulary would be willing to die more cheaply than a regular soldier, and it was hoped that a force of 6,000 native police, well split into detachments, would maintain a semblance of order. As support for this Constabulary, the military authorities expected to maintain greatly depleted garrisons of regular army troops.

How well the idea worked is best shown by the subsequent reduction in regular forces. Moreover, the Insular government would stand the expense of this force, thus providing a pacification force with no cost to the American citizenry.

There was yet another reason to warrant the formation of a native constabulary, that being, as before mentioned, the nature of the work remaining to be accomplished. It was soon apparent that regimental groups of regulars, concentrated in large garrisons, could not maintain law and order throughout the scattered villages, and the very nature of the guerrilla bands precluded the possibility of massed troop movement against them.

Furthermore, the army, as a pacification force, moves under martial law, and is usually unwilling to take the field unless the writ of habeas corpus and other constitutional guarantees are suspended. The presence of the army in the field is tacit admission that the nation is at war. This admission of a state of virtual war was the last thing the Commission desired.

Luke S. Wright held the uncomfortable portfolio of Commissioner of Commerce and Police. Peace was the thing Mr. Wright desired above all things. It was he who proposed the establishment of the Philippine Constabulary, and it came into being as the result of Act 175, Philippine Commission, dated August 18, 1901.

> "An Insular Constabulary is hereby established under the general supervision of the Civil Governor for the purpose of better maintaining peace, law and order in the various Provinces of the Philippines, organized, officered and governed as herein set forth, which shall be known as the Philippines Constabulary."

In deference to the Army, military titles were not adopted for the new force. A group of Inspectors was created to supervise a force not to exceed 150 men for each province, the whole to be under a Chief and four Assistant Chiefs. The pay rates were established much lower than those in effect in the army. Pacification with economy was the watchword.

Imagination gives us a picture of Governor Taft escorting a tall and distinguished Kentuckian into his offices after Act 175 had been made a law. The Southerner was Captain Henry T. Allen, of the 6th United States Cavalry, and Taft could not have picked a better man to whip into line the newly organized corps.

Allen was temperate and severe in a military way, and he was an excellent balance to the rowdy days that brought the Insular Police into being. Jungle of Samar or rue de la Paix in Paris—it was all the same to this gentlemanly aristocrat who had seen so much of the world. He had a code of conduct and he remained with it: it fitted him equally well in jungle or in drawing room.

We are to see Allen later in the course of the fierce fighting in Samar, ploughing his way through high *cogon* grass, three feet to the rear of Captain Cary Crockett, on patrol in very hostile country. He wears the full dress uniform of a brigadier-general; he insists that his officers go into battle clothed as becomes their rank. When grimacing and shouting *pulajans* rise all about the party there in that tangled grass, Captain and Private and Brigadier-General fight for their lives. But always, Allen is the General; if the *pulajans* wish to kill him, there he is, silver stars and all. Allen was a soldier in the grand manner; he was a dashing cavalryman who refused to let the glamour and romance of campaigning ever die.

Allen—"A model soldier; one of the handsomest men I have ever seen," says one of his officers. "Over six feet, with a mustache; a thorough Southern gentleman from Kentucky." The meticulous note again: "We all thought Allen placed too much stress on personal appearance and he seemed to favor officers who had had experience in foreign armies. He was partial to English and Germans."

We have a good portrait of General Allen as he was in 1901. Tall and impressive, the man; with few words and a great personal dignity. Never ruffled, seldom annoyed. Not convivial, seldom cordial, and always courteous. Nor was he a mixer: barbers and tailors came to attend him in his office, and they tiptoed away when they were finished. He was very much against liquor, but was not a fanatic on the subject of temperance.

But one man of all his officers seemed to have a complete lack of awe of the General; of him we shall speak later.

From Constabulary headquarters at Calle Andra, in the

old Spanish Walled City, the General's staff would see him en route to his desk in the morning. He arrived in his own carriage, and each evening the carriage came to return him to his home. On these drives, his wife was always with him.

Allen could be excessively military on occasion, but he was also very much in tune with the times. A story is told of a Volunteer Captain of excellent record who applied for a Captaincy in the Constabulary. He was an excellent soldier indeed—and an exceedingly heavy imbiber of liquor. This Volunteer officer was very anxious to join the Insular Police but, knowing Allen's attitude toward liquor, he had decided to tell the truth and throw up his chance for a commission if Allen mentioned the subject.

As usual, the interview with Allen was severely formal. The General sat there, perusing the Volunteer's papers. The General was impressive and neat—and awe-inspiring. Finally Allen said, "Captain, do you drink intoxicating liquor?"

The Captain rose from his chair in front of the General's big desk. That finished him: he knew that, but he was a very sturdy fellow. "General Allen," he said, glaring across the desk, "I drink all I want, any time I want it. But," he continued, "drunk or sober, I can fight like Hell."

Allen leaned across his desk. "Where are you from, Captain?" he said darkly.

"From Kentucky, sir," the applicant answered.

"Hmm, I thought so," answered Allen, "You will go to Lepanto-Bontoc as a Captain."

The development of the Constabulary was rapid. The restless youth of Manila swarmed to headquarters for commissions. A few mistakes were made in this rushed period of recruiting, but in general the officer list was of high type.

Captain W. S. Goldsborough, Major Wallace C. Taylor,

and Captains Howard Atkinson and Jesse S. Garwood were commissioned Colonels and Assistant Chiefs (Garwood was Major and Assistant Chief), and almost before the public was aware of the existence of the unit, the Constabulary was ready to take the field.

As a quasi-military organization, great care was taken to uniform the Insular Police in some manner that would distinguish them from the army. The material adopted was a soft, cottony fabric, steel-gray in color. *Camano* cloth, it was called. The shoulder ornaments and commissioned ratings were fantastic and wholly unmilitary.

More than a year was to elapse before army hostility had abated sufficiently to permit adoption of a khaki uniform and standardized commissioned ratings for the officers.

We see the new force, then, in the year 1901. They are reaching a tentative hand into the jungle; they are raw, ill-trained, and poorly armed police, facing a mighty epoch of combat. They had 183 officers and 2,417 men at the end of that first year, and they were occupying 94 jungle stations.

The great experiment was in process.

The army was now in garrison, and for a period of many months the entire burden of preserving the peace of the archipelago fell upon the shoulders of the Constabulary. It is a matter of record that from July 4, 1902, until the end of that same year, not a single shot was fired by a soldier of the regular army in the preservation of the peace of the Philippines.

The Constabulary made its presence felt with startling rapidity. We have noted previously that the army strength in the Islands in 1900 had been about 71,000 men, and that their casualties, killed in action, to July 4, 1901, had been 752 men.

The army casualty list *following* the creation of the Constabulary contains illuminating figures. During the period from July 5, 1901, to December 31, 1906, the army loss in men killed was 239. From January 1, 1907, until December 31, 1913, their loss was 23 men. And since January 1, 1914, no American soldier has lost his life in battle in the Philippines.

The reduction in army strength following the birth of the native infantry is as striking. In 1901, the force was reduced to 50,000. In 1902, it had dropped to 27,000; in 1903, to 17,000. In 1904, we had 12,000 regulars in the Philippines. At that figure it remained until the Great War, when a substantial reduction was made, leaving the present army force of less than 5,000 men. These figures do not include native Scout troops.

With these data in mind, we turn back to the organization of the Philippine Constabulary.

With a force of native police preparing to take the field, there remained the troublesome question of their armament. Even the most rabid opponents of the force admitted that the Constabulary could not take the field with police clubs. But by no argument could the General commanding the Philippine Division be convinced that it would be wise to arm the new force with rifles.

There was a certain logic to this objection, for the native force was little more than a dubious experiment. During the months of severe guerrilla warfare, the Filipinos had acquired a deserved reputation for treachery; and there was no reason for believing that the mere recital of a Constabulary oath would make natives loyal American policemen.

One officer of the regulars reported: "By no means arm the Constabulary with smokeless, repeating rifles. Do not arm them with rifles at all. If they are held to black powder shotguns, they will be infinitely less dangerous should they revolt. The smoke of the black powder shells will reveal their positions to army sharp-shooters."

There had been ordered set aside for Constabulary use 1,372 army rifles, and these had been partially issued when the flood of adverse opinion caused General Chaffee to order their recall to the government armories. He objected to the Constabulary bearing the same arm as the regulars, and his position was sustained.

Orders were hastily placed in America for 1,000 Winchester shotguns and a suitable quantity of brass shells. Also ordered were 5,000 Colt revolvers, caliber .45. (This to satisfy the army, who at the time were using .38-caliber revolvers.)[1] No long-range arms were ordered, as it was definitely decided that the Constabulary could not be trusted with a weapon that had a range of more than one hundred yards. So the Insular Police was to take the field with smoke-belching shotguns, against insurgents who had high-velocity, smokeless rifles (the most excellent Mauser of Spain) which had a point-blank range of 600 yards.

And then, there were no shotguns!

The Winchester Arms Company advised that they had no stock of repeating shotguns and were unable to fill the order without great delay. Other equipment and ordnance supplies were similarly lacking, and the men went into action without raincoats, without sufficient shoes or underwear, and with a very deficient weapon.

The administration made frantic scurries to the rejected

1 In this one respect of hand arms, the army gave the P. C. an unconscious advantage. In those days, the .45 revolver was considered cumbersome and unsuited for soldiers. A few years later, when the army fought Moros, they learned the Mohammedans could not be stopped with the .38. The army changed then to the .45, and it has remained the official side arm.

and confiscated arms depots and emerged with single-shot Remington shotguns of an obsolete vintage. One officer recalls them: "As I remember, they were more dangerous to the user than to the target." With the shotguns, they had a few .45-caliber revolvers. One wonders, looking back thirty-five years, when one considers those stores of high-powered Mausers that had been confiscated from Spain. Certainly they remained in the warehouses during this period of arming the Constabulary. Brutal historical facts seem so silly in retrospect. The action was undoubtedly reasonable and logical in 1901.

The situation was so obviously unfair that Sydney Adamson was impelled to remark in *Leslies Weekly* on March 27, 1902:

> "The change in the government from military to civil has robbed the army of supremacy in the Philippines and left it in not the happiest of tempers. The Constabulary is armed only with shotguns and revolvers and events might go hard with them. On account of the class of men who will be enlisted and the caliber of the white men appointed to command, I have little hesitancy in saying that they will be loyal and might as well be armed with carbines at the start. The policy of giving it arms inferior to the ladrones and the insurgents it will engage seems foolish in the extreme. It is calculated to break confidence in their power, to undermine their courage and to give them a feeling of being half trusted. The Remington single-barrel shotgun with black powder ammunition is an easy mark for the smokeless powder rifles of the insurgents and it has a range of but one hundred yards. The success of these brown men at arms under white officers will determine largely the success or failure of the United States in the Philippines."

Meanwhile, the captured Mauser rifles of Spain lay snugly in cosmaline in the government arsenals, and the shotgun force was in the field. Their old Remingtons were belching clouds of sulphurous smoke and the ladrone bands were beginning to know this easily spotted, shoeless force who wore the initials "I.C." on their collars. Knauber was winning his Medal of Valor, and in their out-stations his companions were undergoing constant attack.

The thin-spread station list of the Constabulary during this period inspires awe.

Winfield Scott Grove was patrolling Romblon Island then, with a force of three officers and eighty-six men. He was to have a great career in the Constabulary, and to become the head of its secret service. But in this year of which we write, he was on patrol on Romblon, and knowing Grove as we do, the man must have been in his element there, against odds. He had been a Sergeant-Major in a Colorado Volunteer regiment, and he was very young for his Constabulary rank. He had a deep crease in his head from a *pulajan* bullet on Leyte, a wound that he had not bothered to bandage until long after the fight was over.

"Winnie" is well remembered. How he must have cringed at that name! He was of medium height, solid and chunky, blond and blue-eyed. He was athletic, lively, of quick intelligence and great vitality. "The bulldog type," they said of him. "A swell friend and a dangerous enemy." But, at that, I don't believe he had an enemy in the world.

Sometime later, after his Romblon patrol was finished, we have a view of Grove assigned as new Senior Inspector of Laguna Province. He is tendered a banquet upon arrival to take over his station. The native Governor of the province had been an East Indian, posing as a Filipino, and he had thought well of himself. The banquet had been mostly liquid and inhibitions melted away. When all concerned were pleasantly drunk, this Governor had risen and remarked that he doubted if Grove's ability as a fighting man was commensurate with his actual prowess. "For example," he said, "I don't think the Señor Grove could whip me in a fist fight."

They had one more drink and adjourned the banquet, to retire to the Governor's patio. The Governor was a powerful man and he had studied boxing—these East Indians are surprising fellows sometimes—so the fight was not too uneven.

After a while, Grove came back to the banquet hall to have another drink and servants went out to help the stricken Governor.

The Governor was badly marked up and news of the fracas reached the ears of Taft in Manila. He decided to make an example of Grove, and he sent for the Governor to appear and file charges against the Constabulary Captain. The East Indian executive appeared, one eye closed and his lip badly cut.

"Now," said Taft, "I want you to prefer charges against Grove. This is a scandal." Rumor has it Taft said, "God damned scandal."

Grove was a mighty man—not only in fistcuffs, but in the force of a great personality. And that native Governor was a mighty man too.

"But," said the Governor, "you are in error. Captain Grove is my best friend. He defended me when I was attacked by ruffians. He also got a black eye helping me."

Nobody was deceived, but Governor Taft dropped the matter and Grove was promoted and assigned to secret service duty in Manila. Possibly so Taft could keep an eye on him.

One must approach the campaigns of the Constabulary with an understanding of the fact that the Philippine Archipelago is made up of more than 3,000 islands, ranging in size from tiny coral reefs to the island of Luzon, which has several million inhabitants. Luzon is greater in area than all of Austria.

Dominating the southern scene, as Luzon dominates in the north, is the island of Mindanao, second largest of the group. This mighty island fringes the equator; it is larger than Indiana and three times the area of Belgium.

Some fourteen of the islands of the Philippine Archipelago will be of particular interest in this volume. They are Luzon, with an area of 41,000 square miles; Mindanao, with 38,000 square miles; Samar, with 5,124; Negros, 4,902; Palawan, 4,500; Panay, 4,448; Mindoro, 3,794; Leyte, 2,709; Cebu, 1,694; Masbate, 1,255; and certain smaller islands of the Sulu Archipelago, namely, Jolo, Tawi-Tawi, Siasi, and Bongao. And to these might be added the island of Basilan.

As a basis for comparison of the various land areas of these islands, let us imagine the little State of Rhode Island transported to the Philippines: it would fit nicely into the minor island of Masbate, leaving room on the edges for a large agricultural development.

A description of one of these islands of the archipelago might serve as a picture of them all. The traveler passing them on a steamer today will see on each the same strip of glittering white beach and tangled mangrove swamp that fronts to the sea.

Then the bush begins, broken here and there by the clearings the planters have carved out of it for their rubber and sugar and hemp. The principal islands are spined, in the interior, with a backbone of rising tableland and sheer mountain. These highlands are dense and overgrown with forests of hardwood trees, with *tigbao* and *nipa,* and with bamboo thickets and high waving *cogon* grass.

Jungle Patrol

BIRTH OF THE JUNGLE POLICE

One receives the impression, during a passing cruise in the Islands, that the interior mountains are soft and rounded. Seldom can be seen a spur of jagged rock or the sheer face of a cliff to indicate rugged country. The moutains roll away before the eye, hazy purple and misty with distance; they seem to be gentle, rolling hills. But the rocks and the cliffs and the deep ravines are there. The outlines are blurred with jungle.

The eye of the traveler unknowingly includes a panorama of great jungle and oozing, mottled swamps and game-filled forests and rushing rivers, and wild fields of waving *cogon*. A tremendous, jungle-wrapped country that is uncompromisingly Malay. White man's experiment for a while, perhaps—but never a white man's country. The native population, the climate, the fauna, and the flora all combine in subtle fashion to bar permanent occupation by the Caucasian.

Even today there are too few trails in that silent inner country that is crisscrossed with crocodile rivers; inside, there is no protection from the sting of malaria mosquitos and no let-down or breather if one is to wage successful war against the horrible growth of the jungle.

It is a terrain to try the souls of men. . . .

The total land area policed by the Philippine Constabulary was in excess of 119,000 square miles, an area not appreciably smaller than Italy, and almost identical in size with the whole of the British Isles. To this land measurement add the fact that it was separated into hundreds of islands, each with a peculiar problem, a distinct dialect, and often a new type of mankind. Remember that these islands are swept by difficult currents and often by severe typhoons, and that the coast line is double that of the United States.

Remembering these things, some idea of the beat of the Constabulary can be approximated.

Jungle Patrol

BIRTH OF THE JUNGLE POLICE

In 1901, most of this country was jungle. Much of it is jungle today. When the Constabulary took over the Island patrol service, there was barely one hundred miles of road in the entire archipelago. Army reports show less than five miles of road on Samar, for example. The bush was penetrated by means of trails, which were twisted and overgrown with brush and were often impassable in the wet season.

The rivers of the Philippines are many, but for the most part they are too shallow for navigation in the dry season and they become raging torrents when the rains come in April.

The governmental divisions of the Philippines will be of interest to the serious reader because of the almost constant references to scenes of action by provinces. These provinces are the divisions of the principal islands and they correspond, roughly, to an American state. In the case of the smaller islands, the scene of the narrative can be readily identified by consulting the maps, as none but the largest islands have provincial divisions.

In order to locate specific battle areas with exactness, the provinces of the larger islands are considered here in detail for the convenience of the reader.

On the island of Luzon:

Luzon has some twenty-six provinces. Beginning in the north and continuing to the south, the provincial arrangement is as follows: Ilocos Norte, Cagayan, Abra, Ilocos Sur, Lepanto-Bontoc, Isabela, La Union, Nueva Viscaya, Benguet, Zambales, Pangasinan, Nueva Ecija, Principe, Tarlac, Pampanga, Bulacan, Infanta, Bataan, Rizal, Cavite, Laguna, Batangas, Tayabas, Ambos Camarines, Sorsogon and Albay. All of these provinces may be located on the map in this volume.

BIRTH OF THE JUNGLE POLICE

On the island of Panay:

Panay has three provinces: Antique to the west, Ilo-Ilo on the east coast, and to the north the province of Capiz.

On the island of Negros:

Here we have the island divided longitudinally, with Negros Oriental to the east and the Province of Negros Occidental on the western coast.

On the island of Mindanao:

The subdivisions of Mindanao are very large, and are to a great extent undeveloped in 1937. There have been some changes in the provincial arrangement in recent years. The provinces are eight in number, north to south as follows: Dapitan, Misamis, Agusan, Surigao, and Lanao; and in the south: Zamboanga, Cotabato, and Davao. On Mindanao, there are still 2,000,000 acres unexplored and 5,000,000 acres of standing timber. Less than 12 per cent of the land is under cultivation.

To the south of Mindanao are to be found the smaller islands of the Sulu Archipelago, which are referred to by name in the text and may be readily located. These twin sections of Mindanao and Sulu are known as the Moro Province, and they are the habitat of the Moslems who have so little in common with the Christian Filipinos of the northern islands.

With these brief but essential remarks on the geographical, political, and religious divisions of the Philippine Archipelago, we turn now to the opening reach of the Philippine Constabulary.

That reach was to be inexorable. . . .

The Constabulary blazed their passage through the bush with the bodies of their dead. Their duty was the patrolling of twisting trails over fierce high mountains where there was everlasting and eternal rain; the slogging of dripping swamps, where the big yellow mangrove snakes hung from the limbs; the fording of swift mountain freshets and of slow, gloomy rivers where the red eyes of crocodiles glowed.

As the police of the jungle died, they were replaced by others as hardy and as chill in the face of danger. Theirs was a daily patrol from the little outstations that fringed the dark bush; their lot was to be hacked by bolo and made bloodless by malaria. Sometimes they had no food—sometimes they expended their last cartridge in the face of a bolo rush. And always, that oozing bush overshadowed their lives.

Too many of them had lives of appalling shortness—at twenty or twenty-one, they were snapped off, before they had time to realize their capabilities or accomplish the things they had set out to do.

They built outposts of *nipa* and bamboo, with ground cleared bare of brush and grass for a firing zone. They were simple shacks that sometimes were waiting posts for death. In their stockade walls they made provision for the belching muzzles of their antiquated shotguns. When the dreadful patter of bare feet brought them to arms to meet the swishing blades, they lived fearful short moments that seemed to stretch away to encompass an eternity.

But they endured; they killed and were killed; they patrolled and starved and died of dysentery.

They were insulated from the world with a life constricted to jungle islands that were theirs to defend and explore and to pacify. They ranged those islands, from the equator to the very shadow of Japan.

Chapter iv

SPUTTER OF INSURRECTION

"The Constabulary will patrol the country in their vicinity by night and by day, as frequently as the strength of the force will allow . . ."

—*Constabulary Manual*

IN the later months of 1901, the islands of Samar and Leyte were ignited with a last flame of organized insurgency. The Filipino Generals Lukban and Guevera were there, fighting a final, desperate action, and General Malvar was in the field too, with a force of 3,000 Filipino soldiers.

The insurgent movement had spread to the Province of Batangas, on Luzon Island, and General Bell of the United States regulars began operations in that sector with a force of 2,500 men.

The orders that General Bell issued were ferocious in the extreme, and give a possible hint of the desperation with which the army faced this jungle warfare. "The men will operate in columns of 50 and will thoroughly search each valley, ravine and mountain peak for insurgents and for food and destroy everything outside of towns. All able-bodied men will be killed or captured."

It was decided that insurgents would not be entitled to the honors of war, but could be shot on sight. Protection was given the army lines of telegraphic communication. "The destruction of the wires will call down upon the area where the destruction occurs, complete devastation." It was ordered that all food supplies be concentrated to towns, to be rationed and guarded by American soldiers. This was to starve out the insurgents in the hills.

One of the largest uprootings of civilian population in

American military history was accomplished in Batangas by these troopers of General Bell. One hundred thousand Filipinos were gathered forcibly into concentration camps to be guarded by regulars. A dead line was established, and anyone crossing it without a pass was subject to immediate shooting.

Fortunately, the bark of General Bell was far worse than his bite, for the campaign was conducted with an unusual humanity. One hundred and sixty-three Filipinos were killed by the whole army, 209 were wounded, and 3,626 were persuaded to surrender.

The real renegades remained in the hills, out of the reach of the army movements.

Down in Samar, General Smith was "remembering Balangiga." He is recalled chiefly today for his famous proclamation "to kill every male above the age of ten years on the island of Samar." General Smith was possibly not so bad as he was painted by the subsequent Senatorial investigations which resulted in his retirement from the army. There is much evidence to indicate that he was only following instructions. And possibly all natives above the age of ten *were* dangerous. But there can remain no question that General Smith went to Samar with the self-avowed intention of "making the place a howling wilderness." The memory of Balangiga was very strong.

The picture of our soldiers rioting across the face of Samar is not a pleasant one. They were fighting insurgents, and some of the measures taken there can best have a curtain drawn before them. War, at its best, is a grim business; at its worst, it becomes an horrible business. The army in Samar was no better and no worse than other armies of other nations have been in the course of the subjugation of unwilling, wild people.

These campaigns of Bell and Smith marked the last official operations in the Philippines of the army of the United States as a unit. True, the army performed noteworthy service against the Moros in Mindanao, and they collaborated in the final subjugation of Samar in 1904-1906; but as the mainstay of the peace, they went into the background of the Island scene following the campaigns of 1901-1902.

Only in the Mohammedan country were they to retain control of military affairs for a few years more while the Constabulary was growing to full strength.

General Lukban was captured by the regulars in February of 1902, on Samar, and the entire insurgent army of Guevera surrendered in April of that year. At the same time, General Malvar came in with 3,236 of his forces. The regular army rounded up some 10,000 insurgents in the final drive, all of whom were released after taking the oath of allegiance to the United States.

It was a grand clean-up, and the army deserves all credit.

And so insurgency in the Philippine Islands came to a virtual end. With the passing of the bona fide Filipino Generals, irregular bands began to assume control of the country.

The chilling era was now beginning.

The names of men who will become familiar in this chronicle begin to appear in 1902. The gang of "General" Rios becomes active in the mountains of Tayabas. Rios had been a Captain under Colonel Zurbano of the Filipino army. When the surrender had been negotiated it was not for Rios: he had tasted of power and of the glory of leading men. He sought the hills and there built a formidable semi-religious

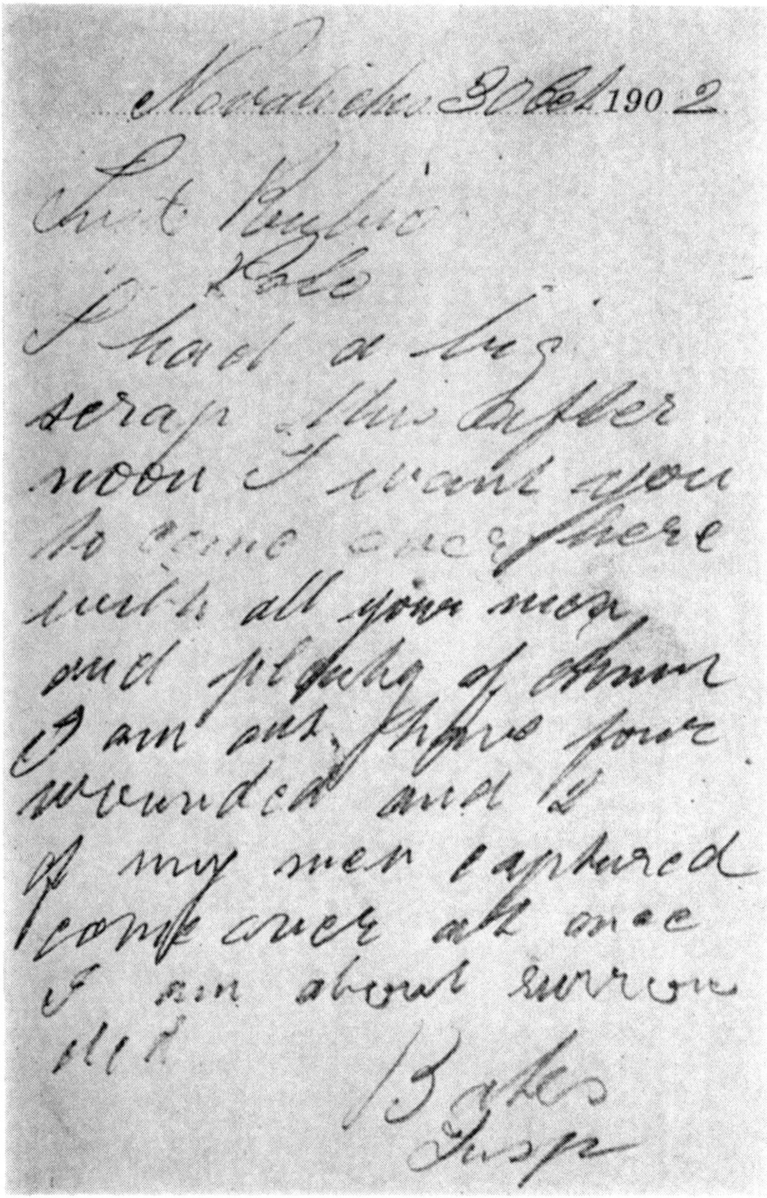

[illegible] 30 Oct 1902

Lieut [illegible]
[illegible]

I had a big scrap this afternoon I want you to come over there with all your men and plenty of ammn I am out, there four wounded and 12 of my men captured come over at once I am about surround ed

Bates
Insp

Despatch sent by Inspector Bates to Lieutenant De Rubio during pursuit of insurgents under Fagin, American deserter.

H. A. C. De Rubio of Headquarters Mounted Detachment

band of outlaws who were plentifully supplied with arms and ammunition.

In Rizal Province, Timeteo Pasay was terrorizing the countryside with a force of guerrillas. They were ragged little men; their tactics were fight and run; they had a facility for dissolving into jungle. The territory of Pasay overlapped that of Faustino Guillermo, whose band in western Rizal had headquarters at Diliman. Diliman was a few miles north of Manila, almost within sight of the office of the perspiring Mr. Taft.

There were others for the Philippine Commission to worry about as they sat in session. From the windows of their offices, they could look across the bay almost to Cavite, where five outlaw bands, totaling 500 men, were in active operation. Among the leaders here was the notorious Felizardo, slippery as an eel and as dangerous as a black panther; with him, Julian Montalon, whose renown among the natives was not measurable. In Sorsogon, not so far away, the fanatical sect of *Anting-Anting* was led by the outlaw chief, Colache.

In 1902, General Allen was engaged with the formation of headquarters troop in Manila. We are unaccustomed to thinking of cavalry operations in the Philippine jungle; but as a matter of record, some of the earliest operations of the corps were undertaken by mounted patrols.

In 1902, the mounted troop, less a detachment in Samar, was assembled at the old Santa Lucia barracks in Manila and its proper training and organization completed. Selected men, representative of the island tribes, were recruited and the officers were carefully graded. It was an honor to belong to the mounted patrol.

At organization, the troop was the Tower of Babel all over again. A dozen dialects were heard in its barracks; one man could not talk to another without an interpreter. Some months were devoted to the ironing out of the language barriers.

Being under the sartorially inclined eye of General Allen, the troop was better dressed and better equipped than other units of the Constabulary. The uniforms were tailor-made and the horse equipment was glistening. An officer writes of the period: "Off duty, in Manila, we climbed out of service uniforms and wore dress white exclusively—caps, shoes, everything—and mostly carried short swagger sticks. I fear we were dandies. I remember that I had thirty white uniforms and often used three suits a day. But as I look back, I wonder if we may not be pardoned that brief respite from killing and campaign."

In the beginning, the troop was mounted upon native ponies—beautiful little animals, but not strong enough to carry man and combat pack through the rice paddies in the rainy season. Also, being stallions, they made so much noise as to make secrecy of operation impossible. To replace them, Chinese ponies were imported; and these being wild Mongolians, a rodeo atmosphere was soon imparted to the barracks. Many men found their way to the hospital in their efforts to subdue their mounts.

The first polo in Manila was played upon these Chinese ponies borrowed from headquarters troop. Out on the old Luneta field, General Allen, Lieutenant Crockett, an artillery officer named Haines, and Langhorne, a cavalryman, started the game in Manila.

The little Chinese horses proved impracticable for military mounts, and an officer was sent to Australia, where some splendid Walers were purchased. These proved entirely satisfactory and were continued in service until the troop was disbanded for economic reasons.

It was a detachment from this mounted patrol that Lieutenant Crockett took into the field with orders to ride the Cavite-Batangas border to prevent the escape of fragments of Malvar's forces who were fleeing the regular army clean-up under General Bell.

The troop had innumerable small skirmishes and were responsible for the capture of more than 200 military rifles and a large store of military stores and hand arms. In combination with troops of the 3rd Infantry under Lieutenants Walker and Sharp, Crockett then led his patrol in indecisive operations against the insurgent chiefs Felizardo and Montalon.

These preliminary skirmishes had one very effective result. They brought to the attention of the administration the capabilities of the Constabulary, and resulted in slightly better equipment for the force.

In the middle of 1902, the Insular Police came into possession of a supply of cast-off .45-caliber Springfield rifles. These were old army guns that had been supplanted by the Krag magazine rifle, but they were better than shotguns. They were single-shot action but they had an effective range of several hundred yards.

With the Springfields, the Constabulary received a supply of Remington rifles, also single-shot, and adequate stores of .45-caliber revolvers and single-barrel shotguns. The official armament of the corps was established at 80 per cent Springfield rifles and 20 percent shotguns. All of the men carried revolvers.

The rifles came into immediate use.

Down on the west coast of the island of Leyte a small

Jungle Patrol

SPUTTER OF INSURRECTION

Constabulary post came into being some few months after the organization of the Insular Police. The tiny post was in an unsettled section; even for the Philippines, in 1902, Leyte was a bad island. We have view of this post, then, at one o'clock of the afternoon of March 27:

A small detachment of fifteen men is occupying a barracks near the seashore. There is no American officer present; the men are in command of Corporal Claudio Circio. They have been there for some weeks, on the edge of trouble. Nothing has happened to break the monotony of garrison duty. Then, this morning, a force of insurgents numbering more than 100, armed with seventy rifles and ten shotguns, and all equipped with bolos, breaks down from the hills and swarms suddenly over the little garrison.

At the moment of attack, six of the fifteen men of the post are outside the barracks. At the first blast of fire from the bush Corporal Circio and Private Paler are killed. The remaining four outside the station seek safety in the *nipa* huts of the town. They are unarmed. Of the four, two privates are overcome with weakness and flee to the bush. The other two leap into the sea and swim far out of rifle range to land again behind the barracks and hasten to aid in its defense.

The detachment is equipped with Remington rifles and a scanty store of ammunition. Under the spirited fire of the eleven defenders, the insurgents engage in a cautious, long-range siege. There is nothing for the beleaguered Constabulary privates to see as they peer through the loop holes to the edge of the jungle. The rifles of the insurgents are smokeless. There is really nothing for those privates to fight for either, unless one counts a few obsolete Remington rifles and a forty-peso station house. But they *do* fight—and manfully, against great odds. Once in a great interval of time, their Remingtons belch smoke at some indistinct figure in the bush.

At three o'clock in the afternoon, after two hours of siege, a lull comes in the firing, and a group of insurgents approach under a white flag. A private of Constabulary, elevated to command, takes the message. They are offered liberty to go free and 100 pesos for each of the rifles if they will surrender the post. This offer to a private, whose pay is twelve pesos each month!

The answer is a volley.

At six o'clock in the evening, the ammunition is exhausted. The force waits silently until it is dark. Then they quietly abandon the post, move along the beach with empty rifles, and find a boat to make good their escape to Ormoc. They return to the post the next morning with reinforcements.

Meanwhile, at nine o'clock of the next morning, before the return of the Constabulary, the insurgents have moved up and fired the barracks. They do not know until the barracks is almost consumed that the detachment has made its escape.

When the supporting party under Lieutenant Crockett arrived on the scene, the insurgents were gone and the barracks a smoldering ruin.

For this defense Privates Gular, Ladrera, Moron, Sampere, Cobacha, Reposar, Clarido, Corillo, Salazar, Padro, and Pelo were awarded the Medal of Valor. The conduct of these Filipino enlisted men, sans white officer, contributed more to public confidence in the Constabulary than any single event of the year.

As the active insurrection had sputtered to a close, a spark of resistance had glowed in the southern provinces of Luzon and on the island of Leyte. General Capili was there on Leyte, and with him the insurgent Colonels Veyra and Pinuranda. It was against these forces that the Constabulary undertook its first major unassisted campaign.

In March, 1902, Colonel Wallace C. Taylor assumed command of all Constabulary operations in this district. The Constabulary, meanwhile, had grown up about him. The roster of the period shows the Constabulary occupying 202 stations with a force of 192 officers and 5,317 enlisted men.

Wallace C. Taylor, in the words of his junior officers, was a "wonderful fighting man," and like Chevalier Bayard, without fear and without reproach. Taylor, with Garwood, was the only fighting Assistant Chief of the early days. He was, (and is) a grand character. A man with "clear-cut aquiline features and a fighting nose," they said. He was the only Colonel not of the regular army, and his district of the Visayan Islands was, in 1901-1907, the liveliest of all. Taylor was the man for the district. Colonel John R. White writes of him, "I yield to him in the matter of field service and experience in fighting outlaws, Moros or anybody else."

In a letter of 1936, Colonel Taylor replied to a request for information with a wealth of material—about other men. Of himself, he writes, "My own service in the 1st Nebraska and in the 39th Infantry was of a lively nature, but not against the odds we had to face against the *pulajans*. The Constabulary service was more like the American Indian campaigns—if you lost there was no support. Your back was against the wall on every occasion. There were many expeditions which tested the vitality of men, many running fights —and a number in which small detachments were worsted. I have always regretted that I did not preserve copies of some of the original reports. I guess we were just young fellows, interested in getting results. . . ."

But the picture of Colonel Taylor that I prefer is one that Cary Crockett gave to me:

"Exceedingly handsome man, very well built, ruddy face, blue eyes, blond hair inclined to curl, hawklike profile, yellowish mustache; utterly fearless. About average height; very trim in appearance and always smartly turned out. He visited my station once in Leyte and we went on a combined pleasure trip and reconnaisance, accompanied by two mounted orderlies only. Accidently, we ran into several hundred *pulajans,* well-organized, well-armed and well-equipped. Taylor galloped in on them, firing his pistol and we, of course, had to follow. They were taken by surprise and fled in all directions. We captured six beautiful horses and a brass band of over twenty pieces. Taylor had set his heart on killing the leader, who wore a fine uniform and carried a big bolo which looked like it was gold-mounted. I had great difficulty in getting Taylor back before they found out there were but four of us and came back on us."

And then a later picture of this fighting Colonel, snapped shortly after that dreadful fight in Samar in which Taylor had one side of his face almost shot away by a *pulajan* bullet. Crockett again:

"'Taylor was two or three days' march in the interior when he was shot. I believe they kept him alive by feeding him raw eggs found in wild chicken nests. I was not present but I saw him a few days later. He insisted upon going riding with me and the wound opened. He nearly bled to death before I could get him to a doctor. I admired him greatly."

This was the man, then, who was in charge of the operations in Leyte. During the campaigns, Taylor had available the headquarters detachment at Tacloban—91 men under

Captain Börseth and Lieutenant Barrett; another garrison at Dolores of 37 men under Lieutenant Waloe; to these were added platoons of the six remaining stations, totaling in all 233 men on station in Leyte. This force was augmented by relief detachments from other provinces to bring Taylor's effectives to approximately 400 rifles. Examination of the records seems to indicate that they were opposed by more than 1,000 insurgent soldiers. Among the insurgent commanders was the American renegade, Harry Long, who held a position of command in the Filipino army.

Split into their usual small detachments, the Constabulary went into immediate action over a large territory. Patrols were sent out under junior officers to scout the terrain and to make contact with the enemy. The army was following this operation with great interest; it was the first test of Constabulary strength and efficiency.

Under Taylor, the Constabulary movements were entirely successful. Patrols captured the insurgent fort at Ormoc, confiscating cannon, horses, and military stores. Between March 30 and June 4, Lieutenant Crockett engaged in eighteen operations against the enemy, accomplishing the killing of Long and the dispersal and capture of insurgent forces.

Long was a deserter from a Volunteer regiment. He had gone over to the natives and set himself up as "military commander" of the island of Leyte. He bushwhacked American forces with a great degree of military skill.

The night attacks of "Colonel" Long were morale-shattering affairs. They would begin with the scream of a sentry which ended in a gurgle as a cat-footed native slit the throat of the outpost. Then would come a fusillade of shots, and then that dreadful patter of bare feet in the dark that marked the advance of the bolomen. The sky would be lighted with long flashes as the Mausers spoke from the bush. The Constabulary camp would be misted with the smoke of black powder ammunition. It would be a slashing pandemonium of noise and clamor and death, with the Constabulary at bay before the bouncing, yelling shapes of the *pulajans*. At dawn, Crockett would take stock; two or three dead men, six or eight wounded, and possibly a man or two missing.

Such was the campaign after Long.

The chase ended with the killing of Long by his own men. His prestige had dwindled under the constant pressure the Constabulary brought to bear, and one morning at dawn he was shot down from behind.

Disturbed conditions of Leyte were largely due to the presence of a strong band under "Papa" Faustino, and a smaller band of outlaws under Juan Tomayo. A lack of sufficient Constabulary force had permitted "Papa" Faustino to build a strong fortification somewhere in the mountains near Ormoc and to inaugurate a formidable coalition of the movement of *pulajanism*. Of *pulajans* we shall speak later. Hibbard, Smith, Adams, and Crockett had engaged these brigand forces in the vicinity of Ormoc without discovering the location of Faustino's fortress. One hundred and twenty-five Constabulary were in the field here for a period of two months, and the fighting was ferocious and incessant.

The Constabulary was beginning to pay a price in lives for each patrol as the fighting grew more grim. At about this time, Lieutenant Neddo moved in with a few men, to Beliran Island.

He was an officer whose great physical courage was tempered by a too great humanity. Had he lived, he would have gone far in the Constabulary. He attempted the pacification of Beliran without bloodshed.

One afternoon, shortly after landing on the island, he made contact with a large force of *pulajans;* he ordered his men to stand at ease and he went forward alone to parley with them and to induce their surrender.

They rushed him and wounded him terribly before his men could rally and beat off the howling, frenzied fanatics. That night his men put to sea, faithful little long-haired Macabebes, carrying their officer with them.

Neddo died before the small boats reached Leyte, and the Macabebes recorded his last words. For the first time in his short life, this final great experience must have frightened Neddo with its nearness, for they said he raised his hacked body and said to them in Spanish, "Don't leave me, my children."

Neddo was an ex-noncommissioned officer of the Army, a slight, silent man who gave his life early for the peace of the Philippines.

But the expeditions all had final result, for, when orders came from Manila on June 4 to suspend hostilities to allow insurgents to surrender, Colonel Taylor was able to report that civil government had again been established in Leyte. On June 15, a few days after the armistice, the insurgent Colonels Veyra and Pinuranda surrendered to the Constabulary at Bay-Bay; and on June 29, at Maasin, General Capili came in with the remainder of his force.

Then in Rizal. . . .

A bandit, Timeteo Pasay, is casting a covetous eye at the arms of the ragged municipal police of the town of Cainta. He bursts from the bush to capture the *Presidente* and the weapons of the police. Then he scurries away to the hills. A tiny detachment of Constabulary leaves station to pursue him through the matted forests. Pasay flits away for a while, drawing the troopers deeper into the unfriendly bush. Then he turns and snaps at the flanks of the police detail.

A man goes down . . . another . . . Pasay gathers up the rifles that have fallen from dead hands and moves deeper into the jungle. One hundred Constabulary then take the field in co-operation with Philippine Scout troops; the campaign ends with the killing of Pasay and the dispersal of his band.

In western Rizal, the resistance consolidates under Faustino Guillermo, who has established headquarters north of Manila. Secret service operatives are sent to spy out his territory. Guillermo, the jungle fox, allows the men of the Information Division to enter his camps. As ragged as the *insurrectos,* these under-cover men live for days in the outlaw camp, as they prepare oral reports for their chief in Manila. About them, the mountaineers sharpen their blades and chant the chorus of the war chants.

One day a council is called. . . .

Guillermo mounts a shelving rock to exhort his followers. The secret service men press close into the ranks to hear what the outlaw chief will say. The ladrone leader has his moment then. "For the traitors who are men of the Constabulary," he calls, "what is the penalty?"

The forest rings with shrill clamor: "The death!"

Not death. *The* death. Days later, a Constabulary patrol parts the jungle fringe and gazes across a silent clearing. The shelving rock is there, but Guillermo and his men are far away. The secret service men have remained, buried to the neck in the center of the crawling, eager ants.

At Novaliches, Guillermo reorganizes and accepts the tenets of the secret society known as the *Katipunan.* Inspector Geronimo takes a small detail against him and is routed with heavy loss. The next day the Constabulary under Geronimo attack again. Guillermo withdraws to the hills.

We next hear of Guillermo on July 16. Late in the evening, attired in blue shirts and *camano* cloth trousers of the Constabulary that he has taken from the defeated detachment of Geronimo, he leads his force into the *barrio* of San Jose, where he completely surprises the Constabulary guard. He captures all of their arms and this time he observes the rules of warfare by releasing his prisoners, unharmed. He is joined here by one malcontent deserter from the Constabulary.

Assistant Chief Jesse S. Garwood then took charge of operations. In appearance Garwood was typical of the old bad men of the West. He was an expert pistol shot, and the possessor of the largest, most droopy mustache in the Constabulary. He was truly a man of arms, later serving as a member of the famed Pennsylvania State Police.

After twenty days of hard marching, the morning of August 14 finds the Constabulary force deep in the forest north of San Juan. The men are marching abreast at wide intervals when contact is made with the enemy. The left flank, twelve men under Inspector Reyes, is struck by a blast of rifle fire. Bolomen flit from the concealing bush and the flank collapses under the rush of cold steel. Guillermo then executes a rapid movement to surprise Inspector Domingo, who is preparing to go into camp with a platoon of ten men. Guillermo strikes, and withdraws after inflicting casualties.

At nine o'clock that same evening he bobs up before Inspector Warren, who is in camp with twenty-one men. The Constabulary soldiers are preparing their evening meal about

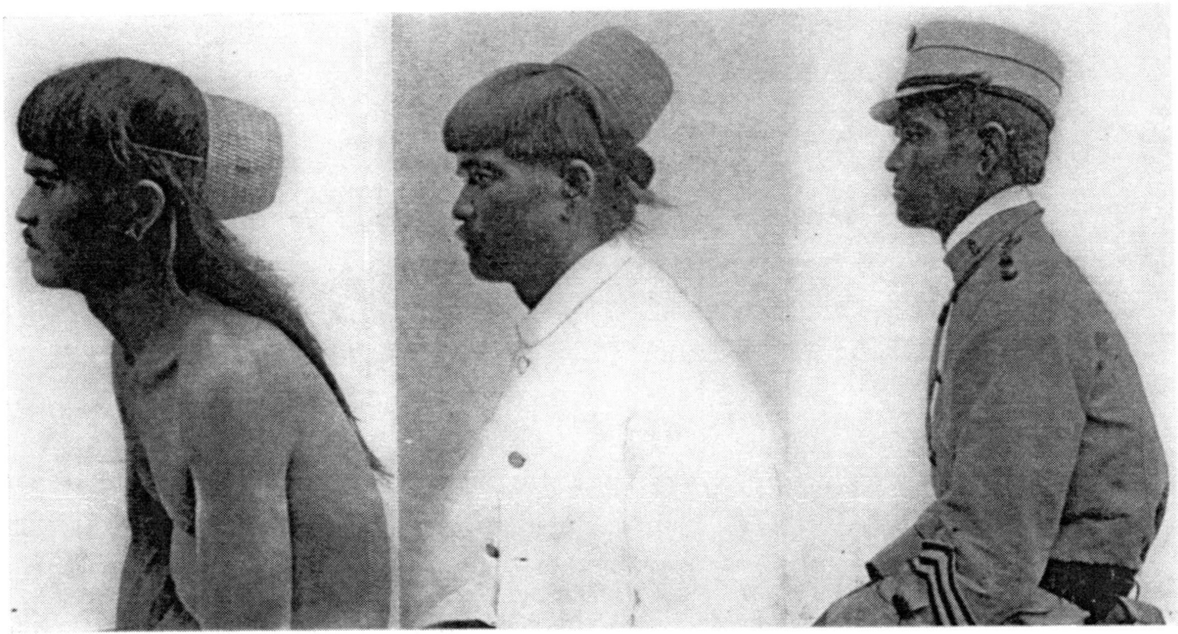

Hillman *Exposed to Civilization* *As a Constabulary Sergeant*

The Evolution of A Constabularyman

Winfield Scott Grove

Jesse S. Garwood

the campfire when the paralyzing rush overwhelms them. Indian-fashion, they take cover in the dark forest and fight for their lives.

Warren and four men are wounded, and at dawn they bury the four privates who went down under that swift stabbing rush of bolomen.

Guillermo is away—not to be heard of again for months.

Jesse S. Garwood, the Assistant Chief who commanded this campaign, is richly worth remembering. He was a study in contrasts and contradictions. He will be recalled as one of the original Assistant Chiefs of Constabulary; he was the most colorful man of a colorful corps.

Garwood was a soldier of fortune, of a type that would have delighted Richard Harding Davis. He was big and blond and handsome, and the ladies thought he was grand; and he liked them all—black, brown, or white, rich or poor.

His memory is rich in stories. See him now, if you will, at a Filipino dance in 1902, as his fellow officers saw him. He has consumed the proper amount of good drinking liquor and is at ease, in a corner, talking things over with a plump, dark wife of the town *Presidente*. He twirls that enormous mustache in beguiling fashion. He leans across, to whisper confidentially in Spanish "Señora, if you will pardon a personal question, I should like to know why a beautiful and cultured lady like yourself should throw herself away on a shriveled-up, wrinkled little cuss like your husband?"

At first the lady bridles, and is properly shocked and very matronly, but as she thinks the whole matter over carefully (all the time under the ardent gaze of those flashing bright eyes), she decides that possibly she *has* thrown herself away and that the Major is a wonderful, observing gentleman.

And then, as Captain Higgins remarks, Garwood would spring the same thing all over again on the first lady of the next town. "Whereever I went in the district, the women would always ask when was Major Garwood coming?"

Gentle with the ladies all the time—that was Garwood. Higgins, who was a Senior Inspector in Garwood's district, speaks again: "I remember a horse-hike with Garwood around Mount Bulasan, taking about two weeks. Every night a new town would give us a reception and a dance. We would take our white uniforms, dirty and wrinkled, from our saddlebags and have them ironed out—no time for washing—and then we would dance all night. At dawn, Garwood and I would mount and ride all day. The next night the same thing. The man thrived on it. Believe me, a man had to be able to take it to trail with Garwood."

Garwood could be solemnly military on occasions. We see him giving an examination for promotion to this same Captain Higgins. Punctiliously, paper by paper, he goes over the questions, twirling his mustache as he waits for Higgins' ready answers. There is an air of great solemnity between the two as Higgins carefully goes through the examination. Higgins has been responsible for those examination questions; he has prepared them in Manila, six months before, as a duty assignment. He knows the answers backwards. But Garwood enjoys giving him the examination in a strictly military manner.

We see Garwood again, in General Allen's office, insisting on supplies he needs for his men. He sets one foot on the General's chair and gestures with a cigar beneath Allen's nose. The formal and ceremonious Allen cringes at such unmilitary behavior, but he can not well ask the Assistant Chief to put away the cigar and come to attention; certainly not an Assistant Chief who makes physical courage somewhat of a romantic gesture. If something was considered dangerous, Garwood would do it, whether it was worth while or not.

The Major was a two-gun man and the show pistol shot of the Constabulary. He wore a big .45 in his belt, and a Smith and Wesson .38 under his arm like a modern gangster. He liked to come into a hostile place and show confidence in his host by taking off his belted gun and hanging it on the wall with his hat. Then he would sit there, debonair and gay—twirling his huge mustache and hoping that somebody would start something. The .38 was handy —just in case.

He was indeed a marvelous pistol shot. He thought he was the best in the corps until Captain Harvey Neville, a very quiet officer, beat him in open competition.

It was Garwood's habit to shoot crows from the top of tall jungle trees with his pistol—or a cigarette from the mouth of an unsuspecting companion. He saved his empty whiskey bottles and was accustomed to treat his staff to rapid-fire exhibitions at these targets lined on the top of a wall. He had a specially trained houseboy whose head served as depository of the apple during Garwood's William Tell performances.

On one occasion, Garwood attended a banquet—his host, Ramon Santos, the Governor of Albay. As the party grew steadily more alcoholic the host said, "Major Garwood, nobody could be as good with a pistol as you claim to be."

Garwood eyed him for a moment. "Claim to be?" he said. "Governor, sure as shooting, you're going to see some shooting."

He whipped the .38 from beneath his arm. At the far corner of the room a row of canned fruit caught his eye. He fired rapidly, with that casual carelessness that made the feat a gesture. Tinned fruit leaped from the shelves—the contents spurting. Then Garwood played his ace. He pointed to the reflection of the governor in a full-length mirror. "Let's assume, Governor," he said conversationally, "that you are a bandit and I am after you. You watch yourself in that mirror." He replaced the gun beneath his arm, and then drew it with the speed of a frontier marshal. The mirror shivered as a bullet caught the reflection of the Governor squarely between the eyes.

It was escapades of this nature that forced Garwood's ultimate retirement from the Constabulary. His humor was too robust, even for the Philippines.

The personality of Garwood popped out at a listener. He had a fund of marvelous stories—to be told in the Garwood manner. He went one day to the Army and Navy Club in Manila to meet a group of officers of the 16th Infantry. Regular army officers sometimes snubbed Constabulary leaders. But Garwood became the center of attraction; he was full of wit and sparkle and they lionized him.

He owned a hundred pair of shoes and he carried them about with him in a special trunk.

We see him another time, en route to Albay to organize the Constabulary in that section. He has a sackful of guns, carried by an orderly; his orders, to organize a Constabulary.

"Just what," he had questioned General Allen solemnly, "does a Constabulary consist of?"

"Use your own judgment," Allen had said. He was unhappy at this levity.

So Garwood devised his own rules and regulations, and one of them was that all applicants should be able to read and write in English or Spanish. One day a Filipino applied.

"Me soldado Americano," patting his chest.

"Ever been a soldier?" Garwood asked.

"Si. Me soldado Español. Guardia Civil. Me *Sargento."*

"Write your name on the paper," suggested Garwood.

The Filipino took pencil in hand and made a fruitless effort to write. Garwood inspected. Not Spanish, certainly; not English; not anything. "No write, no can be *soldado,"* he said. "No write—no fight."

The Filipino pleaded to be made a soldier. Garwood remembered a bandit in the hills who was wanted for murder. "You kill Jose Tinto and I'll make you a *soldado* and give you one hundred pesos."

"Muchas gracias, Señor." The Filipino backed to the door. As far as he was concerned, it was a contract.

Some weeks later, the Filipino came back. *"Me soldado* now; catch one hundred pesos."

"How about Jose?" Garwood asked.

"So sorry, Señor, I no know which fellow Jose, so I bring both."

"Bring them in," Garwood said.

The Filipino went outside to return with a sack containing two severed heads. One was Jose all right. The Filipino became one hundred pesos richer and Garwood recruited a *soldado.*

During the fiscal year 1902-1903, the Constabulary conducted 2,736 expeditions. It was inevitable that so much action should result in occasional severe reverses. The occupation of more than 200 stations with a force of 5,000 men permitted an average station list of but twenty-five men. These isolated posts were under constant attack and were required to hold their positions without reinforcement. At times, half of the station strength were performing regular duty while suffering from severe wounds. There was an ever-present menace of dysentery, cholera, and malaria. The best a patrol leader could expect was 50 per cent of his men in a state of physical health.

Surigao was an experiment. . . .

This town in northern Mindanao had been garrisoned long before the formation of any Constabulary district in the Moro Province. A detachment was there, under command of Captain Clark. They had seen service, for they held in confinement a dangerous prisoner named Encarnacion, who was a fanatical leader of the hillmen. By some means, Encarnacion contrived to escape from the custody of his jailers and had rallied a band in the hills.

The Constabulary were at dinner one evening when Encarnacion came back from the hills. Before the men were aware of any danger, the insurgents had rushed the station and gained possession of the *cuartel* where the arms were racked. The carelessness of Captain Clark caused his death. He was in his quarters, 200 yards from the barracks of his men, when the cries and groans announced the raid. Clark rushed to the scene, armed only with a small derringer pistol, and was cut down and killed.

The attackers came into possession of 148 rifles as a result of this sudden attack. Surigao was a supply station with an accumulation of arms for a section.

As the raid began, the Municipal Treasurer, Captain Kelly, with a few civilians, took refuge in the provincial building and stood off the insurgent attack until a telegram to Cebu brought 700 men of the 11th Infantry, under Colonel Meyer.

Following the raid, Colonel Wallace Taylor threw additional Constabulary into the area and succeeded in rounding up most of the attacking band, with the support of the army force.

The efficiency with which this work was accomplished is best shown by the records of the affair. One hundred and forty-eight rifles were lost in this attack, of which 102 were recovered. Of the 237 men who participated in the raid, 5 were executed, 57 sentenced to prison, 7 acquitted, 25 killed in the action that had followed, and 42 made their escape into the mountains.

At Oas, in the Province of Albay, a similar misfortune befell the Constabulary. For more than a year this section had been harried by forces of guerrillas under the commands of Ola, Sarria, and Toledo. On February 28, 150 men under Magno Revel, a lieutenant of Toledo, fell upon the Constabulary barracks at Oas. The attack resulted in the desertion of twenty Constabulary privates, the killing of those who resisted, and the capture of forty-eight rifles.

Patrols were despatched in immediate pursuit. On March 21, detachments under Captain Linsforth, supported by Lieutenants Fawcett and Grossman, met the band of Toledo in the woods near Buena Vista.

Lieutenant Grossman was in the advance. He pushed a point into the head of a deep, wooded valley. He was a perfect target against that background of green jungle, and the men of Toledo were very close. Grossman went down, shot through the hips with a Mauser bullet at a range of twenty yards. His men fought their way through the ambush and Grossman was carried from the field. He was completely paralyzed, and he died in the hospital at Sorsogon on May 12.

On July 25, men of the 31st Scout Company under Sergeant Nicholas Napolis beat off Ola's force of 310 men near Jovellar. Ola was then attacked in the rear by twenty-five Scouts under Lieutenant Sutherland and an equal number of Constabulary led by Lieutenant Sommers. A running fight raged across the face of the jungle. The Constabulary were out to avenge Grossman, and the aim of the little brown policemen was deadly. Toledo's route was marked by the bodies of his men. In this action, twenty of the outlaws were killed and thirty were found wounded on the field.

So many references have been made to the Philippine Scouts that it becomes necessary to explain this body of soldiery to avoid confusion. The Philippine Scouts are not to be confused with the Constabulary, although they are in certain respects a similar body. The Scouts were (and are) a force of native infantry, officered by white men, the difference being that they were soldiers and not Insular Police. The Scouts are a part of the regular army of the United States and, as such, are subject to Federal law. They were better paid than the Constabulary, and far better equipped and armed. They occupied more permanent garrisons, with comfortable barracks.

The Philippine Scouts, as an organization, antedates the Constabulary. They were organized in February, 1901, and from the first they proved themselves to be most excellent soldiers. Their numerical strength is about equal to that of the Constabulary; in 1903 they had 99 officers and 4,805 men. In common with the Constabulary, they have no regimental formations, having been also designed for small-scale operations.

Although the Constabulary and the Scouts have always been closely co-operative in the conduct of field operations, the personal relationship between the two corps has always been strained. On several occasions the amalgamation of the two groups has been considered, and always the union has been opposed by each faction. In 1903, this condition was aggravated by the order of the Civil Governor directing 29 Scout companies to report for duty under the orders of the Chief of Constabulary and his Assistant Chiefs. General Davis, of the Army, went on record as opposing this shift of authority, stating that it placed army officers serving with the Scouts in a "mortifying position" to serve under the command of the head of Insular Police.

Nevertheless the order of transfer was confirmed, and Scouts and Constabulary engaged in joint operations for almost one year. By these provisions, the active force at the command of the Chief of Constabulary was temporarily increased by some 1,500 men.

All of these detachments were turned over to the Constabulary between February 13, 1903, and July 21 of the same year. The assistance of the native Scout troops, equally familiar with the terrain and the people and much better armed with repeating Krags, was of inestimable help to the Constabulary. It enabled large-scale operations against the last of the insurgent leaders, and permitted reinforcement of stations decimated by casualty and disease.

At Sorsogon, on southern Luzon, Inspector Swann of the Constabulary is at grips with the fanatical society of *Anting-Anting,* headed by the deadly Colache. This leader has swarmed down from the mountains to fall upon and slaughter the city police of Bulusan. Swann makes a forced march across the tortuous hills and the swollen rivers with a slender force of twenty men. After three days of reconnaisance in the hostile region, he finds, not a minor disturbance, but an uprising that blankets the entire coast.

SPUTTER OF INSURRECTION

With his twenty men, he remains in the field, fortifying a small post on the very fringe of the infected area. He sits down there to watch the activity and to await reinforcement. He detaches five of his men, who volunteer, and orders them to report the uprising to the station at San Vicente and to return with all available men.

Incredibly, they win through to San Vicente, but they find no reinforcements at that town. Without hesitation, they retrace their steps through the *cogon* fields and the gloomy forests.

They come to the head of a low, swampy ravine. Ambrosio Fruto, Gabino Dios, and Sergio Dellosa are marching abreast, trigger fingers curled as they scan each tuft of grass. Thirty paces to their rear, Eugenio Faraque and Fernando Filonia are covering their advance.

They approach a shallow creek and the men ahead cross safely. The jungle is very still as they begin the careful ascent of the steep slope that borders the stream. They hear the subdued splash as their comrades behind wade the small river.

A grass clump trembles. . . .

One hundred men of the force of Colache rise from the shelter of the *cogon*. A voice hails the Constabulary privates: "We are all Filipinos, comrades, and should be engaged in a common cause against America. Join with us or not, as you will; you have but to turn your rifles to us to go free."

Fair terms indeed, with the odds 100 to 5! It was Ambrosio Fruto who spoke for his companions; three words he said, that summed up in one short sentence the aims and the code and the creed of the jungle police. "We are *Constabulario*." Then he raised his Springfield rifle and the roar of black powder ammunition mingled with the crack of the smokeless Mausers.

SPUTTER OF INSURRECTION

The Filipino is essentially a fighter who prefers the cold steel. The insurgents fired a scattered volley and then rushed the Constabulary detachment, blades in hand. As the white smoke drifted through the trees, Fruto attempted to break through to the sea that lapped the shore fifty paces to the right. The scene was a mêlée of twisting, slashing figures.

For 300 terrible yards they fought their way. Fruto and Dios fell before the swirling blades. The survivors beat off another rush of the bolomen, but the head of Delossa went spinning from his shoulders under the flick of jungle knives. The two survivors leaped into the sea from a high cliff and swam out to be rescued by friendly fishermen. On the field behind them, seven attacking outlaws occupied the field with the dead Constabulary.

The next morning Swann found the bodies of his dead. From the toes to the top of the head, not one inch of the torsos were unhacked. The arms and legs were disjointed. The stomachs were laid open. Swann gathered his soldiers into sodden sheets and buried them at Bulasan.

Antonio Colache moved on to deserted villages—San Isidro, Santa Barbara, and Santa Cruz. He occupied them briefly; he was constantly on the move, eager to pick off the isolated patrols. Swann, Burton, and Garwood engaged him in dozens of skirmishes, but always he evaded them.

Colache used every native strategy that the jungle had taught him. At night he posted sentries about the Constabulary camps. The bullets of his riflemen spurted the ashes of the campfires and ripped through the *nipa* shelters. His flanks were incessantly active; spearing a careless sentry; disemboweling a straggler who fell behind the columns. The Constabulary were aroused from stiffened sleep to repel the rushes of his bolomen.

In the darkness Colache would move in closely; he would

stake a pig at the edge of the sentry posts and prod it into squealing activity with knife points. It is in that manner that a pig greets the coils of a great winding python. A sentry would approach cautiously, to kill the prowling constrictor.

A shadow would move at his side . . . a blade would crunch on bone.

At dawn, the companions of the sentry would beat the bush and call his name. Then they would move on again, leaving his hacked body, unnoticed and forgotten, where it sprawled in the spreading roots of a mangrove tree.

In April, it was decided to lay a cordon, using native volunteers in co-operation with the Constabulary forces. Four hundred and fifty men were gathered from the loyal villages of Sorsogon, Bacon, and Gubat. By the nineteenth, the cordon was ready. Swann and Garwood divided their slender forces. Swann moved into the hills near Barcelong with twenty men; he was to strike Santa Cruz and move southward. Garwood deployed upon San Isidro and moved north to meet Swann near Santa Barbara. Colache was skulking there.

They met, as planned, but there was no Colache in the toils of the cordon. He had slipped away—deeper into the bush. Lieutenant Burton was detached, with fifteen men, to occupy the deserted town of Santa Barbara. Colache came to life suddenly, and harassed Burton all night in an action that took place in a vast hemp field. Burton maintained his position with serious loss to his command.

The rains came. The action dragged on through the dreary wet month. On April 21 Colache attacked in force, but was repulsed with a loss of ten men. May replaced April, and Swann was carried from the field delirious with malaria and dysentery. Finally, on May 24, Lieutenant Gerona rushed the steep slopes of the mountain retreat where Colache had come to bay and captured the bandit and most of his command.

In this campaign the Constabulary were in the field for fifty-six days, with a force of 153 men. The operations resulted in the capture of 249 prisoners and the killing and wounding of 30.

The most serious menace to the peace of the Philippine Islands in the years 1902 and 1903 was the insurgent General Luciano San Miguel. He was an extremely able soldier and the leader of the *Partido Nacionalista movement.*

The leader of this Nationalist party had been Doctor Dominador Gomez, who was a Filipino Spanish subject married to a Spanish woman. Gomez had at one time been a surgeon in the Spanish army, and after the Spanish American War his influence had extended until he had been elected to the presidency of the *Nacionalista* party, from which position he had edited his reactionary *Los Obreros.* At the height of his influence Gomez had controlled the informal armies of Felizardo, Montalon, and San Miguel. He was arrested for seditious activities and sentenced to four years' imprisonment; and with his passing, leadership was assumed by San Miguel.

About January 15, 1903, San Miguel was elected Supreme Commander of all existing insurgent forces, following his great activity in the wilder parts of the provinces of Bulacan and Rizal. On several occasions he had surprised and destroyed detachments of Constabulary, and his force had grown to a well-disciplined, well-armed army. In Bulacan, in January, he had attacked the command of Captain Warren, and later, in February, the company of Lieutenant Twilley.

On both occasions, the Constabulary had been soundly whipped.

Luciano San Miguel was well equipped for military leadership. He had been one of Aguinaldo's generals, and he had refused to swear allegiance to America. There was a certain honesty in his convictions, and he was respected by the army and Constabulary officers who pursued him. Crockett speaks of him thirty-five years later as a brave man and an efficient soldier.

When San Miguel took to the hills upon the surrender of Aguinaldo, he formed a new *Katipunan* party, which developed into the above-mentioned *Partido Nacionalista* movement, with himself as General-in-Chief in Charge of Military Operations. The document appointing him to command was found on his body when he was killed at Coral-na-Bato. When San Miguel died, the *Nacionalista* party suffered a great loss, for on the field at Coral-na-Bato with San Miguel there fell the General of Brigade Julian Santos and General Benito Santa Ana.

The month of February found every available Constabulary soldier in the Bulacan-Rizal area in the field in an attempt to locate San Miguel and destroy his force. Flying columns of Scouts and Constabulary, each one company strong, were dispatched with orders to contact his army and co-operate in massed attack upon his positions.

Among these forces was the Manila Company A, Philippine Constabulary, under command of Captain Cary Crockett. Crockett had earned for himself a position as one of the premier combat officers of the corps. He had been commissioned an officer on November 12, 1901, and had succeeded to a Captaincy and the command of his company upon the death of Lieutenant Neddo at Beliran Island. His company had been constantly in the field for more than a year and had a record of having never been surprised or defeated, of never abandoning a dead or wounded man, and of having never lost a rifle.

On the night of February 6, Crockett received a message from General Allen:

> "San Miguel reported in vicinity of Bobosco. Costello's force of Scouts is ordered across the mountains and should reach Bobosco tomorrow. A mixed force of Scouts and Constabulary are moving out from Pasig via Antipolo to arrive Bobosco same time. Your company will proceed at once to strike Bobosco from the north and co-operate with other columns."

Crockett was on the march within a few minutes after receiving the orders, and late the next afternoon he came within view of the town of Bobosco. Leaving his force concealed in a ravine, Crockett entered the town with two privates and found the *Presidente* and the town officials huddled in the Town Hall. The head of the *Presidente* was bound with a bloody rag, and upon questioning him Crockett learned that San Miguel, with a large and well-armed force, had arrived at Bobosco the preceding day. His men had cut off the ears of the *Presidente* and beaten him with rattan sticks as a punishment for accepting a civil position under the American government. San Miguel had then crossed the river and taken up a strong position covering the ford and the trails.

There was no sign of other American forces in the vicinity, and it soon became apparent to Crockett that the expected reinforcement had gone astray. He scaled the church tower and surveyed the strong position of the insurgents through his glasses. He knew that he was outnumbered in the ratio of four or five to one, and his men were exhausted from the long forced march, but his orders were plain; and he decided to attack San Miguel.

Having decided upon the assault at dawn the next morning, he dispatched two native runners with a message to Antipolo:

> "To any American Officer,
>
> Arrived Bobosco February 7th without meeting other columns. San Miguel with force estimated at 400 men now entrenched on east side Marquina River opposite town. I will attack at daylight tomorrow. Advise Commanding General, Field Force, Pasig.
>
> "Crockett"

Then, with two Sergeants and one of the Lieutenants, he reconnoitered the insurgent position and discovered a ford some distance below them. He decided to leave Bobosco at three o'clock in the morning, move along a circular route, and attack the insurgent position from the right flank.

At dawn his company was in position at the jungle edge about 800 yards from the insurgent trenches. Before him the ground was open and he had a clear view of the blue shirts and red blanket rolls of the troopers of San Miguel. They were facing *away* from him, intent upon the ford that commanded the town of Bobosco.

Crockett moved swiftly then, with orders to his men to hold fire until the range was 400 yards. Attacked suddenly in the rear in this surprising manner, the insurgents suspected the presence of a large force, and after firing a few shots they retired, leaving numerous dead. Crockett's casualties were two men wounded.

SPUTTER OF INSURRECTION

Crockett's campaigns throughout his seven years' service with the Constabulary were uniformly as successful. He had a genius for attack and a remarkable facility for extricating his men from tight places with small casualties. He was a man of quick and correct decision, and he was absolutely fearless. He was possibly, indeed, the best all-round soldier in the corps.

Certainly the final estimate of any man must come from the officers with whom he served. Colonel Stacey, of whom we shall hear much later, carried with him to France a picture of Cary Crockett in Samar.

Cromwell Stacey was a fine soldier—a regular army man who saw service in four wars. He is well qualified to judge the worth of a fighting man. In 1917 Stacey stood on a reddened field in France beside Crockett as they watched a young artilleryman trying to carry a message through a barrage. Four horses were shot from beneath the messenger. He limped to his feet as his last horse went down and hastened on foot through that Hell of shellfire.

Stacey turned to Crockett then: "Cary, I always said you were the bravest man I ever saw—but damned if that fellow out there hasn't *you* skinned." It had taken Stacey four wars to find a man to compare with Cary Crockett.

As he sat with the writer during one of several conferences, Colonel Stacey thumbed back over his martial memories, seeking for a time when Crockett had displayed the slightest emotion of fear. His face lighted up as he recalled an incident of the trail in Samar, in 1905. I read his expression correctly. "I take it from your expression," I said, "that you have remembered a time when Crockett was afraid?"

"He was scared stiff," Stacey chuckled, "and not that I blame him. There are limits to any man's emotional reserve.

This was in April, 1905. Crockett, Juan Sulse, Wallace Taylor, Captain Green, and I were on a search for Dagujob's headquarters. We were in single file, on a narrow trail, and Crockett was in the lead. He passed a wall of rock that overshadowed the trail and all we saw was a blur of black and a swish as a cobra struck from the edge of the trail. The snake missed and shot across Crockett's shoulder like a spring, and a native First Sergeant killed it with a shotgun. Crockett halted the party and went to sit on a rock. His face was white as a sheet." Stacey chuckled again, "Scared? You're damned right he was—but it took something inhuman to do it."

The effects of the attack on San Miguel were far-reaching. They put the insurgent General on the move, which was precisely what the Constabulary and Scout forces desired. A few moments after the fight was over, the Scout detachment came over the hill and the pursuit was continued without respite for the harried San Miguel. Crockett retired with his exhausted troops, and on February 8, at Coral-na-Bato, Lieutenants Schermerhorn, McIlvaine, Geronimo, and Harris, with 107 men, struck San Miguel in a bloody engagement that lasted for an hour and forty-five minutes.

Here, a young and very blond Georgian gave up his life. Lieutenant Harris had been one of the first appointees from American military schools. He fell at Coral-na-Bato at the beginning of a career.

Lieutenant Schermerhorn, senior officer in this attack, was famous in Constabulary circles as a jungle scout. He is credited with the possession of an uncanny knowledge of the forest and of the ways of wild people. He was a tall frontiersman with a sweeping mustache, and he had great physical endurance. After surviving many bolo rushes without serious injury, he finally came to the end of his active service with the loss of most of one hand as the result of the sweep of a *talibong* blade. He was retired in 1906 for disability.

Following this brush with Schermerhorn, San Miguel took to the jungle, but in March he reassembled his forces at the same place, where he was attacked by the 11th Macabebes, Philippine Scouts, and put to flight with a loss of nineteen killed.

Later in March, a huge cordon was laid along the Pasig River by 400 Scouts and 200 Constabulary. Within its folds large bands of the insurgents and outlaws were trapped and killed.

The Constabulary operations against San Miguel were under the direction of Colonel W. S. Scott, who, before taking office in the Constabulary, had been a Captain of the 1st United States Cavalry. Scott was a kindly gentleman and a fine cavalry officer. He was a rather pallid man with smooth features and a stout figure. He was an excellent administrator, and was slightly bewildered at times by the rapid pace of the youngsters of the Constabulary. He was well liked—even loved—by his junior officers, to whom he assumed the benevolent and kindly attitude of a genial elderly patriarch.

Coral-na-Bato had a fatal fascination for San Miguel, and on March 27 he returned again to this scene of two previous defeats. The 1st and 4th Scout Companies under Reece and Nickersen hastened to engage him. Early in the morning, the force under Lieutenant Nickersen was moving cautiously along the bank of the San Francisco River when the advance point contacted the force of San Miguel in a region that was densely studded with towering bamboo. The outlaws gave way and the Scout force moved in circular flank motion which encompassed the band.

The voice of San Miguel was heard as he discovered this enveloping movement. There was a patter in the long grass and the sound of a few desultory shots. When the point again made contact with the insurgents, it was to find them at refuge behind the stone walls of an old fort. This fortress, built at a bend of the San Francisco River, was composed of but two walls—the other two sides were protected by the river itself.

It was a formidable position, and Lieutenants Reece and Nickersen paused in the shelter of the bamboo to prepare for the assault. A clang of metal sounded along the line as bayonets were fixed. "Attack" . . . the word rolled along the line and 170 men broke from the shelter of the bamboo to rush the stone walls. Halfway across the open field, Lieutenant Reece, who was leading the attack, stumbled as a bullet bit into his body. He recovered, and a moment later the entire force was hand to hand on the walls.

The rattle of the rifles ceased as bayonet opposed bolo. The Scouts swept over the walls, and in a moment the insurgents were in full flight. As the enemy retreated across the river, Lieutenant Nickersen rallied fifteen of his men and crossed in pursuit. His attention was directed almost immediately to several men leading an officer who appeared to be wounded. Nickersen directed his fire at this group. At the first shot from his pistol, one of the group dropped; the rest broke for the brush, leaving the wounded man to make his way alone.

A Sergeant from the 4th Scout Company raised his rifle and advanced upon the weary figure that weaved there groggily in the bright sunlight. The wounded insurgent raised his revolver and fired one wavering shot. Seeing the futility of further resistance, the lone officer raised his hands above his head as he cast his revolver away. "I am San Miguel," he said. A split second later came the retaliatory shot from the rifle of the Sergeant, and General San Miguel pitched forward, shot through the head.

The Scouts suffered fourteen casualties, among them Lieutenant Reece, who was severely wounded. The insurgents left thirty-five dead on the field. But the greatest loss to the insurgent cause that day was in the person of Luciano San Miguel, who lay huddled beneath a bamboo clump with a bullet through his brain.

A few months after the death of San Miguel, the town of Vigan, Province of Ilocos Sur, was captured and held for three days by an insurgent force, which included a number of Constabulary mutineers. A general uprising of the whole of northern Luzon threatened.

Colonel Scott ordered out Crockett's company and a Macabebe Scout company under Nickersen, and they proceeded north on a coast guard cutter to land twenty-five miles south of Candon. Within an hour after disembarking, Crockett, who knew the region, had established contact with the insurgents. The two companies attacked and defeated a larger force, recovering arms, U. S. mailbags and government property. Captains Crockett and Nickersen made a night attack following this victory, and succeeded in rounding up the entire movement. The surrender was made to Captain Nickersen. This operation was the final blow to insurgency in the vicinity of Vigan.

With the passing of San Miguel, the final heartbeat of the Philippine Insurrection sounded. His death was followed by the surrender of many minor leaders, and never again was the United States to encounter resistance from any legitimate leader. San Miguel must be rated a sincere insurgent and not a bandit. The leaders who followed him were bandits.

The remaining "generals" came in slowly to be amnestied or executed or imprisoned, and one of them, Salvador, was not apprehended until 1911. Another, "General" Noriel, was convicted of murder and executed in 1915, for a crime committed in 1903.

The Constabulary was facing ahead now to bitter resistance, which was little more than bushwhacking at the hands of bandits and ladrones who developed from the collapse of the insurgent movement.

As the Insular Police had advanced through the last phases of the insurgent campaigns, the organization was affected by the passage of much favorable legislation. On March 6, 1902, an Act of the Commission had provided for the creation of supply officers to supervise the branch commissaries that came into existence. Three months later, on June 9, the Commission authorized an increase in strength of the force to a maximum of 5,000 men.

Most important legislation of all was the passage of Act 568, dated December 23, 1902. As a Christmas present, the Constabulary officers received the dignity of military titles. Although we have hitherto called them Lieutenants, Captains, etc., to avoid confusion to the reader, actually, until December, 1902, they had all been Inspectors. Now the old First-Class Inspectors became Captains in fact; Second-Class Inspectors, First Lieutenants; Third-Class Inspectors, Second Lieutenants. The Fourth Class Inspectors had a title created for them; they became Third Lieutenants. To the end, the Constabulary retained that title as the lowest rank of commissioned service.

In the first quarter of 1903, the Commission passed additional legislation providing for a fifth Assistant Chief of Constabulary; established disciplinary measures for punishment of torturers, insubordinates, men absent without leave, and for negligence on sentry duty, and made provision for suitable quarters for officers on station in Manila.

In April, a sixth Assistant Chief was authorized, and the number of officers in each grade was increased by five. In June, four Majors were authorized, and the pay of ten Captains was increased to not more than $1,800, and of ten First Lieutenants to not more than $1,200.

Meanwhile, a system of competitive examination for Constabulary officers had been drafted by Captain E. R. Higgins on August 16, 1902. The subjects and their weight ratings were: spelling, 5; arithmetic, 5; letter writing, 10; penmanship, 5; paper work, 15; history, geography, and civil government, 10; Constabulary drill regulations, 15; Spanish and native dialects, 10; general fitness for service, 25. Of the 138 officers examined, 23 failed.

At about this time, too, the Constabulary adopted the khaki uniform, with standard shoulder insignia and the red epaulets that remained their costume to the end.

There is a little town on the island of Catanduanes, which is off the coast of Albay. Virac is its name, and it is possibly much the same today as it was thirty-four years ago when Harrison O. Fletcher took his Constabulary to barracks there. It must have been a lonesome outpost for the few Constabulary soldiers who stared away across its strip of white sand to the swirling currents of the Gulf of Lagonoy.

SPUTTER OF INSURRECTION

It was the afternoon of August 13, 1902, and Fletcher and his men had returned from a routine patrol into the hills. The Lieutenant was seated in his quarters at the seashore, watching the steamer *Dos Hermanos,* which was at anchor two hundred yards from the surf line. As he scanned it idly, a chorus of shouts came from the small vessel; then a puff of smoke and the rattle of rifle fire. A moment later, figures flitted across the decks and cries of "Kill them—kill them all"—floated across the water.

Fletcher called two of his men and set out in a small boat to investigate the disturbance. As he pushed off from the sand he saw figures on the deck gaze in his direction; there was a rush of men to the bow and a frantic heaving of the anchor chain.

We turn to the reports of what Fletcher accomplished that day. There is something in this story that reminds one of the Texas Ranger tales of "one man for one riot."

The reports indicate that Fletcher stood up in his small boat and called out in Spanish to drop the anchor or he would open fire. For answer, a rifle bullet splashed in the water beside the bow of the dinghy. The Constabulary laid a blast of rifle fire along the deck then—if two riflemen operating single-shot rifles can deliver a blast. The men on the deck continued to haul at the anchor cable, and Fletcher's men laid down their rifles and paddled in under the quarter of the steamer. The *Dos Hermanos* was beginning to move in the direction of the open sea.

Covered by the fire of his two men, Fletcher seized a trailing rope and hoisted himself to the deck of the steamer. When he had cleared a place with revolver fire, his two men mounted to stand beside him.

Those three valiants fought off and cowed forty-five mutineers on the deck of that vessel! As the battle raged across the deck the shouts and cries of the barricaded passengers and ship's officers added to the confusion. The mutineers soon had enough of Fletcher's accurate, close-range fire, and they leaped over the rail into the sea to be captured by Constabulary who lined the shore in front of the barracks.

Fletcher found the Chief Engineer of the *Dos Hermanos* dead from ten bolo cuts; the Steward lifeless on the galley floor; the Captain, Second Mate, Second Engineer, and two passengers badly cut with bolos. It had been the intention of the crew to kill passengers and officers and make away with 15,000 pesos that was in the ship strongbox. Fletcher, with two men, accomplished the capture of thirty-five of the mutineers, killed three, wounded five, and permitted the escape of but two.

When news of this action reached Manila, the Medal of Valor was awarded Lieutenant Fletcher and Privates Victorio Penalosa and Ruperto Nolla.

Shortly after this *Dos Hermanos* affair, Lieutenant Fletcher engaged in one of the most dramatic single-hand combats that is recorded in thirty-five years of battle.

The old-timers call it the "affair of the bicycle". . . .

It was eight o'clock in the evening, and Fletcher was slowly pedaling a bicycle along the lonely, palm-lined road between Camalig and Guinobatan in Albay. He was heavily armed for a special mission, having a .38-caliber revolver and a Krag rifle slung at his shoulders. His road turned sharply and angled away through the gloom of a coconut estate.

He was rushed in the dark by twenty-three bolomen.

As the patter of feet sounded, Fletcher dismounted and hurled the bicycle in the path of the charging men. The rush was delayed, partially, but not before he had received several serious wounds. He wheeled in the very shadow of the slashing blades, drew his revolver, and fired until the click announced an empty weapon. The bolomen, who had paused momentarily in the face of that accurate revolver fire, re-formed and came back at Fletcher again as he clawed frantically for the carbine. It came into play just in time and under its rapid fire the outlaws wavered and retreated into the shadows.

As Fletcher crouched there in the road with blood streaming from his wounds, an almost incredible incident obtruded to save his life. He heard, in the distance, the voice of an American soldier from the regular regiment on station at Daraga, calling to determine who was being attacked.

Fletcher turned his eyes away from those lurking shadows that ringed him and made a quick decision. He wagered his life on the belief that the lone American soldier would understand Spanish. He called in a loud voice, "Bring up your detachment and attack."

The single American in the distance not only understood the Spanish language, but was quick to appraise the situation. He advanced on the double, giving orders, "Load magazines, right by squads, double time." As he led his imaginary platoon to the rescue, the outlaws faded into the bush. Behind them, they left five dead and four wounded, as evidence of the accurate, under-pressure fire of Fletcher.

The year 1903 drew to a close. . . .

General Allen was destined to go far in his chosen profession of arms. It was to be his privilege, sixteen years later, to be summoned to the command of the American Army of Occupation in Germany. He was to see, in the later years of his military commands, thousands of young Americans die on the fields of France.

But as he sat in his office in Manila in 1903, to recapitulate the record of the corps he had brought into being, it may have been the most satisfying point of his long career. For he was living in a personal era. Achievement was a personal thing in those days. Death was a personal thing, too, in that small group of companions in arms. His corps had weathered a year of incessant fighting; the wonder was that the casualty lists were not longer.

But there was a brighter side to that record of a year and a half of patrol. His force had gathered in 3,019 firearms that would be no more a menace to the peace. His men had conducted 8,087 expeditions against the enemy on patrols of law and order that reached the impressive total of 332,923 miles. The men with red epaulets had killed 1,859 disturbers of the peace, and had captured 5,539 outlaws.

The Constabulary was fulfilling the expectations of the Philippine Commission, and more. They were on patrol from Ilocos in the north to the Mohammedan islands in the south. Theirs had been a life of daily contact with the enemy; such had been their idea of the duties of a policeman. More than 8,000 expeditions in eighteen months of existence!

Personalities were emerging from the fabric of the corps —gallant, half-starved personalities who were frayed by jungle. In their skimpy uniforms of gray they had been, unknowingly, creatures of a great romance as they had led their barefooted patrols deep into the bush. Now they were erect in khaki, and morale had come to the force, and a great confidence in their arms.

Already they had learned that in these desperate patrols, brute strength to brute strength could not be. They laid the foundation, in 1902 and 1903, for their code of arms. "To be outnumbered, always; to be outfought, never."

They substituted an acquired jungle cunning and an individual excellence, to make up for their lack of numerical strength and their single-shot rifles. The personnel of the force was combed relentlessly in this period to eliminate the unfit, the indecisive, and the ones who lacked in courage.

And their reach into the jungle grew longer.

Chapter v

THE "POPES"

"Men most useful for scouting are those having a greater power of observation than the ordinary man; such men should be selected and trained to develop their greatest use for field service."
—*Constabulary Manual*

THE Filipino has a flair for secrecy. . . .

The *Katipunan* had been the original expression of their urge for underground organization, and it had had its beginnings in Spanish times. Organized by one Del Pilar, the brother-in-law of Rizal, the spread of the society had been so rapid that by 1896 it appears that the *Grand Council de Katipunan* had a roster of at least 200,000 memberships. The society had been designed to precipitate armed revolt against Spain, and its members were bound by the fearful *pacto de sangre*—the covenant of blood.

With the numerous revolts of the *Katipunan* against Spanish authority, we are not concerned in this volume. It is sufficient to say that these uprisings were put down with great bloodshed and by mass deportation to the Spanish penal colonies in Africa.

What does concern us is the existence of the active core of the society which carried over to the time of the American occupation and which was largely responsible for the organized resistance of the Filipino army. The *Katipunan* was also responsible, in later years, for the growth of many of the numerous quasi-religious sects whose members terrorized Samar, Leyte, and Luzon. There is evidence, too, of *Katipunan* influence in the raids of the organized bandits who caused the Constabulary so much concern in the interval following the insurrection.

THE "POPES"

The *Katipunan* had been largely responsible for the Diliman gang under Faustino Guillermo, captured at Coral-na-Bato. The Mendegorin gang of Zambales and Bataan had also been under the influence of the society. This band, previously not mentioned, had been the first to be eliminated in the regular army clean-ups. Santiago Alvarez, President of the *Partido Nacionalista* movement, had been a confirmed *Katupinero;* Sakay, one of the vice-presidents, and later founder of his own movement, had been active in the mountains, issuing *Katipunan* commissions.

When the active insurrection had dwindled to a close, there remained throughout the archipelago, scattered but active *Katipunan* lodges composed of flaming zealots who were unwilling to admit that their cause was lost. The original principles of the *Katipunan* fell apart during this disorganized period and the bolder of the members turned to the dubiously patriotic arts of banditry.

This fragment of the *Katipunan* grew again in strength as it slowly spread across the face of the Islands; it revived the flaming hatred for the United States and entered into secret rebellion with an organization of the mountaineers.

The ignorant hillmen of the provinces were well disposed to listen to the inflammatory suggestions of the *Katipunan* agents, for ill times had befallen the Islands. The country had been laid waste by war and was in the throes of a painfully slow reconstruction. The situation might be compared to the demoralization in the South following the Civil War. Many natural calamities had been inflicted upon the land. The cattle and *carabao* were dying of anthrax or were being driven away by the robber bands; 100,000 men and women were dead of cholera. Famine, locusts, rice worms, and flood and drought had completed the rout of the Filipino farmer.

It was the old, old situation of the urban people exploiting a rural neighbor who could ill afford to be exploited. Manila merchants sought to corner the short rice crop, and the Philippine Commission had been forced to appropriate $1,000,000 to stabilize the rice price and import essential quantities of this commodity from Saigon. It was a period of starvation, with each man taking what he could from a weaker neighbor.

The Filipino *tao* of the farmlands is dependent for existence upon his *carabao*. These great, placid, long-horned water buffalo are the tractor of the Philippines. They are the transport of the farmer and the cultivator of the rice paddies. Their broad backs form the cornerstone of success or failure for the Filipino planter. The problem of the shortage of the *carabao* was in great measure the root of the serious guerrilla warfare which took command of the Islands for more than a decade.

In 1901, the supply of *carabao* had been reduced by disease to but one-tenth of the normal number. Their cost rose from the usual twenty pesos to more than two hundred pesos. With their shortage, the rice crop had dwindled to less than 25 per cent of the normal yield.

It is understandable, then, that the Constabulary interspersed fighting with a great activity in the recovery of stolen and strayed animals. In many respects, their services in this not glamorous duty were more valuable than their prowess with the Springfield rifle. When the *carabao* became scarce, they became precious.

The guerrilla warfare was almost a war waged for possession of *carabao*. The big work animals became objects to be acquired by the strongest, and it is small wonder that many honest farmers joined the revolutionary movements to give their work animals the protection of an armed band.

THE "POPES"

The peaceable farmer was in an unenviable position. As a nonmember of the secret groups, he was subject to constant raid and extortion; as a member, he was harried by army and Constabulary. As a nonmember, he was also subject to reprisal if he gave any aid to the forces of the United States who were seeking the extermination of the outlaw bands.

It was in Cavite and Batangas that the secret sects first began to appear in strength. There, an elaborate schedule of punishments were drawn up to impede the activity of American troopers. Noncombatant natives were served with this warning:

> For taking office under the American government — death.
> For giving information to Americans — cutting off the lips.
> For guiding American troops — cutting tendons of the feet.
> For giving supplies to Americans — crushing fingers with rocks.

Nonmembers were subjected to rigorous taxation, and a system of collection was established through a chain of dread *inabans,* or agents, who rivaled the Russian secret police in ferocity and ability to strike terror.

Into this disorganization and economic starvation the *Katipunan* sent capable and oratorical leaders who soon inflamed the ignorant mountaineers with vague promises of a millenium that would succeed American rule. Men opened their veins to mix blood in the *pacto de sangre,* and religious ceremonial dominated the jungle meetings.

Simple hillmen awoke to find themselves "Generals" of ragged, shouting armies of fanatics. . . .

As outgrowths of this *Katipunan,* other societies appeared.

In Cavite, Sakay set himself up as "President of the Filipino Republic," and selected Carreon as his "Vice-President." With him was Montalon, one of the most feared of the insurgents, and Felizardo, who was one of the most bloodthirsty villains of the period. Montalon held the portfolio of "Lieutenant-General of the Army of Liberation," and Felizardo was one of his chief murderers and torturers.

In the north, in Luzon, Felipe Salvador burst into prominence as head of the *Santa Iglesia,* or Holy Church Organization. He posed as a prophet, and affected the long hair of a Biblical saint. His emissaries made use of the old Spanish Weather Bureau to circulate among ignorant farmers predictions of floods and typhoons. His prestige became enormous and adherents flocked to his banner. It was Salvador who had engineered the attack on the Constabulary garrison at San Jose, detailed earlier in this account. He had then retired to the swamps of the Candaba, where he was in the year 1904.

Salvador had been a deserter from the army of General Aguinaldo, and he had set up a system of brigandage in the Province of Nueva Ecija, where he had been an adept plunderer for more than ten years. He was believed by all of his followers to have supernatural powers, including that of invulnerability. Even when he was captured, his followers believed that he would escape or that he would have a second life after death.

This fanatical society of *Santa Iglesia* was at least semi-religious. It had been started by one Gabino, who was captured and shot by the Spaniards in 1893. From its original designation as *Gabinistas,* it had been changed, in 1894, to *Santa Iglesia.*

They carried crosses and rosaries, and had a ritual adapted from certain features of the Catholic ceremony. Otherwise, there was no similarity. Salvador preached socialism, practiced polygamy, and promised that the land and other desirables would be distributed among his followers when the government was overthrown. He preached that then would come a great fire to destroy all unbelievers; and after the fire, a rain of gold and jewels upon the faithful. He said that a wooden club would turn into a rifle if used bravely enough—but that lack of faith would leave it still a wooden club. Salvador had headquarters on Mount Arayat, from which his followers believed his spirit occasionally took flight for interview and personal visit with the Divine Powers.

When Salvador went into one of his prolonged trances he was accustomed to demand female companionship, which was eagerly furnished by the fathers of the more desirable girls of his flock. The long-haired man was accustomed to circulate freely among his subjects, in the main treating them well and offering up public prayers in the market places.

We know little of the *Santa Iglesia;* today we recall them as a brotherhood that dealt in death. The records of their complicated religious rituals are vague, and they are best remembered by the soldiers in the Philippine jungle who heard their chant of "San Pedro" and saw them leap to attack with the swishing of long crescent blades. The history of the Philippines is filled with records of their bravery; we have many pictures of individual fanatics dying for the belief that their wooden club would turn into a rifle.

They were the Crusaders of the Philippines; they were valiant, wholly futile, religious murderers.

San Felipe Salvador had secured 100 Mauser rifles in an attack upon the Spaniards at Dagupan in 1898. With a force thus well-armed, he had been appointed a Colonel under Aguinaldo in 1899, and when official hostilities came to a close he had refused to surrender. His subsequent activities show that this decision was not prompted by patriotism but by a desire for great personal power.

In 1902 he had been arrested by the Constabulary, before he had become a "Pope" and a religious leader, but he had escaped and set himself up on Mount Arayat to enter a new phase of his activities.

In Leyte, the Constabulary had to contend with "Papa" Faustino Ablena, who had a stronghold in the mountains near Ormoc. Faustino was a man in the fifties; he was fifty-three years old when killed. He, too, had a reputation dating from Spanish times, for he had been arrested in 1887 for organizing the *Dios-Dios* organization and was sentenced to a prison term in San Ramon at Zamboanga.

With the coming of the new régime to the Islands, Faustino saw the opportunity for "Popehood," and began to work upon the credulous people of the foothills to invest himself with an aura of the supernatural. He signed himself "Señor Jesus y Maria," and began the distribution of charms, love potions, and religious trumperies with a tone of paganism. Lieutenant O'Conner's attack upon his citadel will be discussed later in this volume.

On the island of Negros, a bad cholera epidemic was seized as an excuse for activity on the part of the *Babaylanes* or *Montestas*. Under the guidance of "Papa" Isio, with one Dalmacio as an aide, this sect threatened the lowlands and inflamed the population with a proclamation which stated that the cholera was caused by the Americans having poisoned the wells.

Many months were required to settle the cases of Delmacio and "Papa" Isio. Isio had been a *tao* laborer and a herder of cattle. In 1880 he had fled his position as cattle tender for a wealthy Spanish family, after the killing of a citizen under circumstances which pointed clearly to murder. In the mountains he had gathered a band of men, and having a persuasive personality, he soon was able to begin his forays to the lowlands. In 1896 he had made a disastrous attack upon Magallon and had been repulsed by the Spanish Guardia Civil with a loss of fifty killed. Then he had attacked Cabancalan with better success.

In 1898 Spain turned over the government of the Islands to the Filipinos in an effort to impede American progress. Isio was called in from the mountains by Spanish officials, and wined and dined and given a pretentious uniform.

From that series of interviews Isio emerged a "Pope." He was a source of great annoyance to the American government and was the object of army and Constabulary patrols for years. He was erroneously reported killed in 1905. Of his final end we shall speak later.

On Samar Island, "Pope" Pablo had consolidated factions of the *Dios-Dios,* which was to develop into pulajanism, discussed in the next chapter. The *Dios-Dios* had been under Anugar, who now relinquished the leadership to the "Pope." Pablo was the religious head; Pedro de la Cruz rose to become *Jefe Superior de Operacions;* Isidro Pompac, better known as Otoy, became *Segundo Jefe de Operacions.*

Pablo's unofficial army was completed by the sub-chiefs Aguilar, Anugar, Jose Jerna, Vicente Picardal, and the murderous Amongo, who is better known as Teducduc.

Nor was the business of "Holiness" confined to the male sex. On one occasion four "Virgin Marys" were in jail in the Philippines. The activities of the two women "Saints," Margarita Pullio and Catalina Purical, should be briefly mentioned. These ladies were busily engaged for several years in the manufacture and distribution of *anting-antings,* or charms against bullets. The ladies decided, on September 16, 1903, to relinquish life in the mountains and surrender to American authority.

During the period of "Papal" resistance, the following "Messiahs," in addition to those mentioned above, were eliminated by Constabulary and regulars: "Papa" Fernandez in Laguna; "King" Apo in Pampanga, and many lesser "Saints" who came less under the bright glare of publicity. The period produced three "Jesus Christs" and one "God Almighty"— all of whom occupied Constabulary jails during this period of religious terror.

Of all of the "Popes," Rios the *Tulisan* was possibly the most famous. The Province of Tayabas, which had had its full share of religious fanaticism, spawned him. He was a religious curiosity: a queer combination of philosopher, bandit, and psychologist. His *Tulisan* movement was an offshoot of the *Colorum* sect, which had established a "New Jerusalem" on Mount San Cristobal near the dividing line of the provinces of Batangas, Tayabas, and Laguna.

Ruperto Rios represented himself as a god, and he found small difficulty in establishing a sect of fervid worshippers about him. So well did he succeed that he organized what he called the "Exterior Municipal Government" of the Philippine Islands (for revenue only), and he set up a pompous régime that bristled with titles. He had a liking for uniformed attendants, and he was fully aware of the value of a sonorous title in the impressing of a simple hill folk. He made it possible, almost literally, for everyone to be a general.

He promoted his men so rapidly that he had about him one captain-general, one lieutenant-general, twenty-five major-generals, fifty brigadier-generals, colonels and majors by the dozen, and lieutenants by the hundred. Over the whole was Rios, as "Viceroy of the Philippine Archipelago."

In time Rios became dissatisfied with earthly titles, and announced himself the "Son of God" and the "Deliverer of the Philippines," placed on earth for a divine mission. As proof of the miraculous power with which God had seen fit to endow him, he dispensed *anting-antings* that were guaranteed to make the wearer invulnerable to attack.

The ceremonial nature of the *Tulisman* movement is best indicated by the paraphernalia captured by Captain Murphy on March 8 near Infanta. Among this equipment was a box upon which was painted the word *"Independencia."* The followers of Rios believed that when they had proven worthy, the Prophet would open the box and this mysterious thing, independence, would come forth to bless them.

Captain Murphy desired to hear more of this marvelous *independencia* that was contained in the box of "Pope" Rios. His men brought to him a slightly wounded *Tulisan,* who was interrogated.

"Si, Señor; in the box it was, but by now it has flown away."

"Flown away?"

"Si, Señor Capitan—to the 'Pope,' to be enclosed again in another box."

The fanatic rolled his glistening eyes as he drank in the thought of the approach of the millenium. "When *independencia* flies from the box, there will be no labor, Señor, and no jails and no taxes."

Fair enough, thinks Captain Murphy. But the "Pope" has promised even more than this. "And there shall be," the native adds, "no more *Constabulario."*

No labor, no jails, no taxes, and no Constabulary — a principle worthy of war.

Of Rios we shall hear more.

The various religious and political sects that engaged the attention of the Constabulary would fill a long page. Among them may be mentioned the *Dios-Dios,* the *Colorados,* the *Cruz-Cruz,* the *Santa Iglesia,* the *Tulisan,* the *Cazadoes,* the *Colorums,* the *Santo Ninos,* the *Guardia de Honor,* the *Hermanos del Tercio Orden,* the *Babaylanes,* and the fanatical *Anting-Anting* which had been headed by Colache.

In general, these organizations had a close similarity to one another, and their origin could be traced to the original *Katipunan.* The recruits were malcontents, *insurrectos,* and more often, murderous banditti. In a few instances, the organizations had a sincere purpose. In general, they were against all law, preying equally upon Americans and their own countrymen.

Here is a proclamation signed by one of these organizers, to indicate the manner in which the sects went about the business of undermining the civil government:

> "The Presidente,
> Municipality of Minglanilla,
>
> "I, being the envoy of the powerful God to arrange this Province, undoubtedly I will fulfill what the Almighty God has disposed, because the time for the liberty of the Philippine Islands has already sounded and will so happen. I have sent communications to all towns, asking if they intend to take part in the general outbreak. In case favorable, send here guns and revolvers in your charge.
>
> "Anastacio de la Cruz
> "The First Teacher"

The religious emissaries constantly invaded the ranks of

the Constabulary with suggestions of mutiny and desertion and the killing of officers. The rewards were alluring: places of high command in the hill parties, with commissions as "Generals."

In the light of the temptations offered, the loyalty of these ill-paid privates was almost unbelievable. During the first two years of its existence, the Constabulary lost but fifty-nine men by desertion.

With this understanding of the wave of religious fanaticism that had swept across the country, it may be seen that the nature of the resistance had changed. No longer were the fighting forces in the Philippines concerned so greatly with political groups; it was pseudo-religion that had succeeded as the force behind the combat.

The third battle phase was roaring to life. . . .

Chapter vi

THE PULAJANS

> "Detachments marching where attack from bolomen is to be apprehended should have the column absolutely closed up so that in the event of attack the men will be able to render mutual support . . ."
>
> —*Constabulary Manual*

NEAR Dolores, in the Province of Tayabas, there is a sacred rock to which the hillmen have gone for worship for many generations. As part of the simple ritual of their services, they wore red sashes. In the early days, it had been the simple religion of a simple, childlike people.

A people called *pulajans*. . . .

We have reserved for final consideration, that native movement of *pulajanism,* which was the most formidable of all and, in a manner, a coalition of all of the sects. These red-garbed mountaineers, with white flowing capes and crescent blades, were contributory to one of the most ferocious eras of guerrilla warfare that our arms were to experience. Not even the Indian campaigns of the old West, fought in open country, could compare with the rushing, jungle-shielded tactics of the *pulajans*. For ferocity in battle, possibly only the Moros were their equals.

When the army completed what they thought was the pacification of Samar, the Constabulary took over responsibility for the law and order of the island. It was a demand beyond their limited strength, for Samar, with its *pulajans,* became too hot for the thin-laid patrols of the jungle police.

Today, we think of the word *pulajan* as descriptive of a clan of mountain bandits. Actually, the *pulajans,* in their beginning, were no organization at all. They obtained the name from their distinctive dress, the word *pula* in Visayan meaning red. The *pulajans* had been simple highlanders, cultivating the mountain clearings in peace. Without exception, the hillmen of the Philippines are timid and peaceful, unless prodded.

The *pulajans* were the *Babaylan* of Negros, the *Colorum* of Batangas and Tayabas, the *Santa Iglesia* of Nueva Ecija and Bulacan, and the *Guardia de Honor* of Pangasinan and the Ilocos.

The *pulajans* had been forced into banditry by a combination of vicious circumstances. From time immemorial the hillman has raised his crops of rice and hemp, and borne them laboriously on his shoulders to the sea settlements for barter and trade. The hillman is a retiring creature and not too well understood by his sophisticated cousins of the cities. His religion has no complicated ritual; it involves the flowers and the trees and the inanimate figures of Nature about him. He knows little of the arts of trade and finance and commerce.

Too often the hillman is made the victim of shrewd merchants of the coast. Such had been the experience of the *pulajans*. The report of Governor-General Smith, as late as 1908, confirms that the *pulajan* was a victim of circumstances beyond his control. The Governor-General writes:

> "The *pulajan* is not a robber or a thief by nature — quite the contrary. He had his little *late* of hemp on the side of the mountain, and breaking out his *picul* of the product he carried it, hank by hank, for miles and miles over the almost impassible mountain trails, to the nearest town or *barrio*. There, he offered it for sale and if he refused the price tendered, which was generally not more than half the value, he soon found himself arrested on a trumped-up charge and without hemp or money."

THE PULAJANS

The original *pulajan* trouble that flared in Samar and Leyte had its beginnings in financial transactions between the highlander and the coast merchant. The robbing tactics of the latter brought on retaliatory *pulajan* raids. The raids increased in frequency and severity, and the glow flamed to a blaze with a hitherto peaceful people suddenly aware of their capabilities and in command of the island. In turn, the lowlanders became the victims.

As the *pulajan* movement grew in strength, it became impregnated with a tone of religious ritual and frenzy. It came, too, under the influence of unscrupulous native leaders who saw in the hillmen the agents for their own personal advancement. Samar and Leyte became filled with crusading "Popes" who were self-appointed "Messiahs" and who soon impressed their influence upon the childlike mountaineers. The seeds sown by the "Popes" began to bear fruit, and the *pulajans* now became militant crusaders with a developed hatred to law and order, and an homicidal intent toward their own countrymen in the lowlands. Their red tribal costumes became bedecked with white crosses, and the raids grew into ferocious forays that were out of all proportion to the original grievances of the mountaineers.

From sympathetic figures, the *pulajans* developed into inhuman monsters.

Probably no white man can write with authenticity of the *pulajan* movement and of its meanings. What peculiar form of religious fanaticism caused these natives to gather at designated places for prayer and preparation for battle cannot be told with certainty. But gather they did, groups numbering sometimes as many as a thousand fighting men. The number of organized *pulajans* eventually to assume virtual charge of the island of Samar has been estimated at from 7,000 to 10,000 warriors.

THE PULAJANS

Their weapon was a heavy, crescent-shaped bolo with which they could decapitate a man at a blow. Their battle preparations consisted of bottles of holy oil, prayer books, consecrated *anting-antings,* and other religious paraphernalia. Their mode of attack was a massed bolo rush. Their battle cry was that dreadful "*Tad-Tad*" which means "Chop to pieces," and they moved into action behind waving banners.

From a military viewpoint, their tactics were unsound, as they gave no thought to casualties. They were contemptuous of death, and they rushed without thought of position or the possibility of encountering enfilading rifle fire.

They could be stopped by a determined stand of accurate riflemen if the odds were not too great. Often, the odds *were* too great and it resulted in the death of every soldier who faced them. When the *pulajans* once got to close quarters with their great knives, massacre was the result.

The men who have survived their charges write that never can they forget that scream of "*Tad-Tad*" and that patter of bare feet in the jungle that announced the beginning of a *pulajan* charge.

And so, an originally simple hill people came under the influence of "Popes" and exploiters to develop into unsympathetic and bloodthirsty bandits.

There is the incident of the town of Taurian. This village came under *pulajan* attack and twenty-six peaceable natives were slaughtered by their own countrymen. There is another incident of the *barrio* of Cantaguio to remind us of the bloody raids. Here the *pulajans* came in force at daylight on July 10, 1904. They killed the Lieutenant of Police and several others and proceeded to the *tribunal* (town hall), where they seized the local *teniente,* or municipal president, an appointee of the United States government.

He was conducted to the plaza by a cordon of red-shirted men and the American flag was fashioned into a turban for his head. This was saturated with kerosene and lighted. Surrounded by his fellow townsmen, the *teniente* was burned as an example to his fellows of the dangers of accepting office under the Americans.

As the torch flamed, one Julian Caducay, the leader of the *pulajans,* commented to onlookers upon the dangers of serving that flag. "Call upon the flag you have adopted," he said to the stricken *teniente,* "to protect you *now.*"

The band then cut off the lips of the *teniente,* burned the village, and carried off fifty of the inhabitants. These latter were principally girls who were to be the concubines of the bandits.

For this atrocity, Caducay was hunted down and captured to be hanged.

It was a wave that swept the Philippine Islands. . . .

A frantic, kaleidoscopic tide that drew together *pulajan* and *Santa Iglesia* and *Cazadore* and Mohammedan Moro alike—to spew them out on the crest of the bloody froth that lapped the combat years. It was a hemorrhage of clawing, homicidal tribesmen who had little in common save that desperate urge that spelled resistance to man-made law.

It was a jungle—alive now, with fangs of steel—pouring its pagans and its Christians and its Mohammedans against the lonesome patrols. The *"Tad-Tad"* of the *pulajan* and the *"La ilahu-il-la'l-lahu"* of the Moro were blurred and almost silenced in that thunder of sound that signaled the conquest of the jungle . . . that dreadful chop of blade on living flesh; that blast of the Krags, and that whistle of spears in flight.

In that jungle are the enduring footprints of the men who brought law to the Philippines. The large footprints of men of great stature who served in the American jungle patrol. Some were regular army; some were Scouts; the rest were officers of Constabulary. In their daily duty in that frightful bush they were fulfilling a prophecy that had been expressed in the magazine *Truth* in an issue of 1899:

> "We wish you much joy in your Islands,
> Which you have so easily won.
> But the troublesome part of the business,
> Has only, we fear, just begun.
> You will find how extremely ungrateful,
> Your new fellow subjects can be,
> Compelling you even to shoot them,
> Before they consent to be free."

But the men of the jungle patrols were greater than the colonial urge that brought them into existence. They showed, by everyday valor and by a splendid fortitude, that they were to be set apart from ordinary men. They were magnificently careless men—many of them bearing the names of able fighting families of America. Lieutenant C. E. Boone, of the 14th Ilocanos, must have understood that jungle, for he was a grandnephew of the immortal Daniel who had blazed the way Westward in America.

We have a view of Daniel's grandnephew in Samar, hiking the trail with Cromwell Stacey. The two officers enter a region where a strange bright red fruit resembling an apple and a tomato hangs from the branches beside the trail. Stacey speaks, "Boone, I wonder if those things are poisonous?"

"I wonder," Boone answers.

A few hours later Boone speaks again to Stacey. "About those red things," he says. "They are not good to eat—and they are not poison. I ate two to find out."

And Boone, in a measure, was speaking for all of the members of the jungle patrol. It was their manner of doing things; the only manner they knew. To investigate—to find out—to break through the secrets of the jungle. Quite often they died before they reached that last full limit of their endurance. The force they commanded was not adequate to the qualities of leadership these men possessed. They proved that, in later years, in other wars.

Today, officers in white uniforms attend teas at post receptions where these muddy riflemen of three decades ago fought for their lives.

The men who conquered *pulajan* and Moro are not contributory to a book on the jungle campaigns. They *are* the book.

Chapter vii

BANDITRY

"Once a ladrone camp is located, there must be no hesitancy in attacking vigorously. The best time for attack is at, or before daylight and during rainy weather. . . ."

—*Constabulary Manual*

WE have seen the development of Isio into a "Pope" through the kindly offices of the retiring Spanish government. He was more than sixty years of age when he came to the height of his power as the "Messiah" of the *Babaylanes* of Negros. His whole lifetime had been one of active warfare and insurrection. Even as early as 1897 he had been a powerful figure in the Philippine scene.

The American 6th Infantry had started the first serious operations against Isio in 1902. With their replacement by the Constabulary, the search for him continued under Major Orwig; it was to end in his capture in 1905.

In 1902, "Papa" Isio was the acknowledged King of Negros and the leader of the mountaineers. His forays against army, Scouts, and Constabulary had greatly enhanced his reputation as a supernatural creature who could not be harmed by bullets. Isio had capitalized upon this credulity; few of his men were not equipped with *anting-antings* scrawled with the name of the "Pope" and promising immunity from rifle fire. With this band of dangerous fanatics he ranged the inner country, swooping down with favorable opportunity to augment his rifle and ammunition supply at the expense of isolated patrols.

In this manner he came into La Castellana one morning

in 1902, killing several Scout soldiers on station there and escaping to the hills with twenty repeating rifles.

On October 10 of that year a young Constabulary officer celebrated his twenty-third birthday by taking command of the district of Bacolod, on Negros; sixteen stations, 200 men, and some thousands of square miles of jungle. This was John R. White, who was no neophyte in this matter of fighting, in spite of his years. He had seen service with the Greek Foreign Legion, with the army of the United States in Manila, and was later to see service with that same army in France. In 1902 he was red-headed, freckled, very impulsive and eager, and champing to get about the business of the extermination of Isio and his gang.

Captain White (now Colonel, retired) was one of the finest officers to be produced by the Constabulary. His record embraces every facet of Insular Police service, including the wardenship of Iwahig penal colony. He was respected and admired and was one of the best liked officers of the corps. He was an administrator and he was a fighting man. He retired from the field of Bud Dajo, badly wounded, with a Medal of Valor.

In 1902, he was chasing *pulajans.*

The island of Negros has an area of slightly less than 5,000 square miles. For convenience, call it the size of New Jersey. It has in the interior a backbone of high mountain ranges that rise in places to 8,000 feet. In the southern portion near the Binalbangan River, Isio had his headquarters in the mountains. Farther north, on the slopes of Mount Canlaon, his collaborator in crime, Dalmacio, had an equally well-established sphere of action.

Dalmacio was a remnant from the prehistoric past of the Philippines. His mother had been a Negrito, and this tribe of broad-headed, broad-nosed little black mystery people had been the original settlers of the Islands. Centuries before the coming of the white men, the Negritos had been driven by later arrivals of Malays into the most inaccessible mountain regions. There they had developed a jungle cunning and a close community with the creatures of the wild.

Dalmacio, the half-caste Negrito, had inherited the best traits of two races of the East. From the Malay, he had received the heritage of a great personal valor, and it was fortified by the jungle knowledge of the little black Negritos.

White began his Negros operations with a fruitless series of operations against the wily Isio. The old *Babaylane* proved as elusive as a butterfly, and considerably more dangerous. It appeared for a while that the ancient bushman might surrender to Colonel Kennon of the 6th Infantry, but this hope faded and the Constabulary began the long jungle search.

Macabong, the "Pope's" headquarters in the mountains, was located and burned, but the troops who had made a long night march in the hope of surprising Isio found the citadel empty.

The "Pope" had merely retreated to his fortified post at Mansalanao, south of Canlaon, where he had a strong position on the summit of a lofty peak. Here, his position was thought to be impregnable, with trenches on the mountain top overlooking the trail and with an area cleared away to permit rifle fire upon an attacking force. The trail to this fortress was trapped with *belatics,* which are spears lashed to yielding saplings and constructed to be thrown with great force when a careless foot trips the vine trigger. In the trail, too, were carefully-placed bamboo splinters, pointed and poisoned to wreck havoc on barefooted soldiers.

On the twenty-fourth of May, White set out to attack this fortress. He had a force of seventeen Constabulary soldiers!

BANDITRY

On the afternoon of the third day of march his little party was deep in the mountains on the slopes of the hills that were crowned by the "Pope's" seat at Mansalanao.

As the detail moved along a narrow trail they heard the sound of voices and the pad of men's feet in the trail. They took cover in the grass. After an interval, a line of native *cargadores* approached, and they intercepted it to find it to be one of the "Pope's" trade convoys.

By this fortunate circumstance they captured unwilling guides who were forced ahead to warn of spear traps in the trail. Many times in the course of the trek up the mountain-side the guide stepped aside to avoid cunningly concealed bamboo slivers or to indicate a spear poised to impale a man at the edge of the bush. Without casualty, they made their way through the "Pope's" first line of defense and late on the afternoon of the twenty-sixth White brought his detail to a halt at the foot of a hogback.

Three thousand feet above them was the position of Isio. It was wreathed in mist and was forbidding in the distance.

As the young officer faced ahead to survey that jungle-clad mountain slope, he was mulling a decision that came to every commander of a Constabulary patrol. That was the fact that, miles deep in the jungle, with a very insufficient force, he was about to attack a vastly stronger foeman. White sat on a rotten log and surveyed the *Babaylane* position through his field glasses. It looked bad; they had every advantage of position and outnumbered him greatly.

He fell in his men after a hasty meal of rice and sausage. He began the ascent of Mansalanao.

In places White found the trail so steep that steps had been cut to make possible the ascent of the *cargadores* of Isio. The captives were ahead in the trail and the *belatics* became more numerous. The Constabulary Sergeant would step forward, as they beckoned, to slash the fastenings of a spear trap with his bolo. They moved ahead steadily and saw no outposts of natives.

At three o'clock, after a hard climb, they made first contact with the enemy. At a bend in the trail, they came suddenly upon a patrol of the *Babaylanes*, and White raised his non-regulation Winchester .44 and took a hasty snap shot. Immediately the air was filled with jagged boulders from above.

The big stones rolled down into the path of the Constabulary, who took cover in fissures of the rock until the avalanche had passed. The *Babaylanes* were in two lines of trenches above them and White ordered his men to pick off all heads that showed above the first parapet.

As the shower of stones slackened, the Constabulary rose and rushed the first trench, to come under the fire of Mausers and Remingtons at a range of 300 feet. The mountaineers' volleys were high and did no damage, but their work with the native weapons was more effective. Every man of the charging Constabulary was wounded by the spray of spears and the renewed shower of rocks.

In the shelter of the first trench the Constabulary waited for an hour while a continuous stream of spears passed over their heads. Under this hail of missiles, White held his position with but five men. Twelve of his men were under cover below, reluctant to face the swishing spears. The Lieutenant mounted a rock in view of his men, and called to each by name, all the time being under fire from above. Bullets passed through his sleeves and a stone smashed into his face, but he remained there until his men beneath summoned their courage and advanced to the first trench.

Once assembled there, the Constabulary rushed the remaining *Babaylane* position, and nightfall found them in possession of Mansalanao, with "Pope" Isio in full flight. White burned the fort and destroyed many tons of rice, sugar, and other food supplies. In the last rush, he had succeeded in killing one of the most dangerous war leaders of the "Pope." This man, Aguacilcito, had been shot through the head in the final advance.

This successful completion of a fight that should have been impossible resulted in a cessation of *Babaylane* activities in White's sector. Isio moved deeper into the mountains.

A few months after the assault on Mansalanao, the island of Negros became ridden with a cholera epidemic. The Constabulary vainly threw cordons into the infected territory in an effort to halt the spread of the disease, but all sanitation measures were useless and 25 per cent of the population died within a space of a few months.

Dalmacio, the Negrito, then came into the picture.

He issued a proclamation stating that the cholera was the result of the Americans poisoning the wells. Men flocked to him, and he began to harass the outlying Constabulary posts. Dalmacio would not risk a pitched battle; his activities were sudden nocturnal raids, followed by periods of inactivity in the mountains.

White determined to draw him out of the mountains, and he sent small squads of men to engage the Negrito and withdraw, giving an impression of defeat. Under these tactics Dalmacio grew bolder, and when his force had grown to 200 men, with many rifles and one cannon, he determined to risk a battle with the thirty-odd Constabulary under Captain White.

White advanced to meet him with twenty-two of his soldiers. The forces met at daylight in the suburbs of the *barrio* of Murcia and a desultory running fight followed, with Dalmacio withdrawing before the police advance.

After an hour of this, Dalmacio found a position to his liking and took a stand on the banks of the Caliban River, where a high bank had been previously prepared with a system of trenches. White moved in close and came under rifle fire. When within 200 yards of the *Babaylane* trenches, Dalmacio's cannon boomed and a homemade projectile composed of rusty nails, bolts, and a kerosene tin, dropped at White's feet.

The Constabulary Captain soon decided that he would not be able to rush the position across the open country. The position of Dalmacio was on the far side of the river. White detached a Corporal and a few men and gave orders for a continuous fire upon the *Babaylane* trenches. Then, with Lieutenant Smith and the remainder of his men, White made a wide circuit and crossed the river below the scene of the fight. Creeping on hands and knees through the high *tigbao* and *cogon,* he came suddenly upon the *Babaylane* trenches from the rear.

As they rose from the grass, they were within fifty feet of Dalmacio's men. The *Babaylanes* turned and fired at point-blank range. The first volley killed two men of the attacking force, and the rifle White held was knocked from his hands by the impact of a bullet. He retrieved it, and the two officers, each armed with a repeating rifle, pumped bullets into the natives.

It was the element of complete surprise that saved the lives of these two American officers, who were far in advance of their men. Had the Babaylanes chosen to rush with bolos, the issue would have been different.

Instead, they broke and scattered through the grass, and the fight became a chase with Constabulary and fanatic pouring across the hills in individual combat. Captain White went in pursuit of two fleeing *Babaylanes*. The first he shot through the body and he passed the threshing form as he continued in pursuit of the second. Then he had a snap shot through the tall grass at the second figure and his quarry went down, shot through the legs.

As White approached, the wounded *Babaylane* pulled trigger at point-blank range, but the revolver missed fire. The outlaw then drew a bolo and began to squirm through the grass to the officer. White raised his rifle, reluctant to kill the wounded man. The *Babaylane* decided that further resistance was useless. He threw down his weapon and was taken prisoner.

It was Dalmacio, the Negrito, greatest of the Negros chiefs.

Elsewhere on Negros, "Papa" Isio was in full flight, and was never able to rally his full fighting strength. He remained at large for several years, with badly diminished prestige; and at length, after sixteen years of outlawry, he was surprised and captured by Lieutenant J. S. Mohler. The "Pope" was sentenced to prison and in these later years his warden was John R. White, who had campaigned against him in Negros.

The old "Pope" could not stand captivity and after a short imprisonment he died in his cell, and his ambitious program of a "Kingdom of Negros" collapsed with him. The leadership of his mountain clan was assumed by one Eugenio Alcachupas, who continued the grotesque religious rites and ceremonies. But the old spirit of the "Pope" was lacking, and with the passing of Isio, the showman, peace came to Negros.

The year 1903 had seen the extermination of the bands

of San Miguel in Bulacan and Rizal; of Modesto Joaquin in Pampanga; of Roman Mandalag in Zambales and Pangasinan; of Timeteo Pasay and the Felix brothers in Rizal; of Dalmacio and Rufo in western Negros; of Flores in Misamis; of Anugar in Samar; of Concepcion in Surigao; of Colache in Sorsogon, and of Encarnacion in Tayabas.

The year also saw the gradual shading of insurgent activity into a period of out-and-out brigandage. Many of the bands that had been led by bona fide insurgent leaders now came out under lesser leaders on forays for plunder and revenge. Sometimes they shielded their lawless activities with organizations of high-sounding political significance; at all times they were more dangerous and more difficult to exterminate than had been the insurgents who had preceded them.

The bandits were elusive as quicksilver. The chilling uncertainty of guerrilla bush warfare became the rule. The outlaw bands grew smaller and more nomadic. They wandered across the face of the archipelago, losing a few men here, gathering recruits there. Always they were on the move, and always they had designs on the rifles of the Constabulary patrols.

Firearms became scarce after the insurgent round-ups. They commanded a high price in money and lifeblood. The combats grew small in scope and more frequent. The action developed into swift stabbing frays, with twenty or thirty bolomen involved.

Unique methods of warfare were developed to combat this new emergency. The cordon came into being, a long, loose, and widely spaced formation of Constabulary, closing in a wide circle to entrap outlaws like rabbits.

Not always was the cordon successful.

The nature of the formation prevented a concentration of

men. At Das Marinas, 1,200 Constabulary and Scouts collaborated to form a living barrier across the bush. The circle began to close. In the shelter of the jungle, the outlaws formed a compact spearhead and drove through the thin-spread line. Headless bodies were left in their wake and the scarred, chopped torsos of men who had been hacked to death by the flashing bolos.

At Das Marinas the outlaws showed their contempt for the cordon by mutilating two secret service men inside the circle of soldiers and hanging them for the troopers to find.

As the hysterical wave of banditry swept the country, it was inevitable that its virus should affect the Constabulary. Privates on lonesome patrol duty began to see their countrymen wandering the islands, taking what they wished, with their fill of women and goods and money. A few of the privates succumbed to the constant temptation and abandoned their stations for the bush. At Polangui, six privates wearied of the rigors of police service and reverted to outlaw station. Lieutenant Jahn took a detail to apprehend them, and they had time, for ten years, to meditate in prison upon the evils of disloyalty to their military obligation.

The fighting had been treacherous. Lieutenant Kiely had been one of the first victims of the new brand of back-stabbing warfare. In Capiz, with a small patrol of four men, he was searching for outlaws. He entered a house where he was informed that a sick man was dying. The supposedly sick man sprang to his feet as the Lieutenant entered and severed an artery in Kiely's wrist with a sweep of a bolo. An enlisted man inside the house with the Lieutenant was also severely wounded. The assailant leaped from the window and was shot by the two men of the patrol who had waited outside.

Kiely endeavored to bind up the wound that was spurting his life away. He was carried to Calivo and there on the day following he died from loss of blood.

Sometimes the outlaw bands consolidated briefly and made attack in force. In northern Neuva Ecija, the *Santa Iglesia* leader, San Felipe Salvador, gathered his following and attacked the Constabulary at San Ramon. He waited until midnight, and then he sent 100 attackers under "Capitan" Berong.

The Constabulary detachment, numbering thirty men, were surprised, with but a moment to prepare a defense before the outlaws were upon them. The sentry was killed at the first volley and the *Santa Iglesia* succeed in reaching the breastworks before the Constabulary were at stand-to.

The police poured from their quarters to engage the outlaws hand to hand. For more than two hours the battle raged in front of the Constabulary barracks.

Captain Wakeley stood at the stockade, the remnants of his force about him. A bullet thudded and the Constabulary Captain was lined with flame as thirteen cartridges in his belt exploded with a roar. Wakeley staggered from the force of the explosions and turned to find himself face to face with the bounding figure of "Capitan" Berong.

The outlaw leader raised his voice above the din. "We are here by appointment," he called. "Keep your promises, kill your commanding officer." Wakeley dropped him with his last bullet and moved into the fracas swinging a rifle he lifted from the hand of the dead man. The Constabulary swarmed over the barricade and drove the outlaws to the bush.

After that pandemonium of hacking, slashing, jabbing battle, only eleven of the Constabulary were on their feet. Wakeley took stock; eight of his men were dead and three more were severely wounded. The waist and hips of the Captain were scored with powder marks and he had a bullet in his side.

In this action, thirty Constabulary engaged 100 *Santa Iglesia* in close combat. They succeeded in killing eleven of the bandits, wounding four, and capturing three.

Wakeley and six of his men were awarded Medals of Valor. Here, as in other similar engagements, the superior discipline of the Constabulary was all that saved them from annihilation.

The leader Simeon Ola was now heard from again.

Ola had been an insurgent Major under General Belarmine, the Tagalog leader. Ola was a small man and his rank had given him delusions of grandeur. He had taken advantage of his Majority to treat his neighbors with great severity, with the result that he had become a pariah in his native village of Guinobatan.

Among the natives who had suffered at the hands of Ola was one Circilio Juacain, a Chinese half-caste who had now risen to affluence to become *Presidente* of Guinobatan. He had immediately driven Ola into the hills, where the ex-Major affiliated first with Jose Roldan and later with the formidable Toledo.

In time, Ola became such an efficient murderer that he was elevated to the command of the consolidated band. His depredations became such that three Assistant Chiefs of Constabulary successively took the field against him. Baker tried to capture him; then Garwood, the mighty warrior, took a hand.

Garwood's actions were typical of the man. He went alone into the mountains, to the camp of Ola, to try to persuade the outlaw leader to surrender. He was led blindfolded, for six hours, to the retreat in the hills. There, ringed by 1,500 outlaws, he made the camp ring with laughter as he sat around the fire with Ola, telling robust stories and eating stewed grasshoppers with the bandit chief. After a pleasant interval, under conditions made to order for Garwood, Ola agreed to surrender.

Garwood then resumed cartridge belt and revolver, which he had discarded as usual, to show confidence, and mounting his horse, was led back to his waiting men.

But in this case, Garwood's mighty reputation had been to his disadvantage. Ola had had no intention of surrendering. He had used Garwood's visit as propaganda to increase his force and his position. The prestige of Garwood was enormous among the hill people; his visit to Ola in the hills was hailed far and wide as a recognition of the outlaw by a great American soldier. Recruits flocked to Ola.

It was H. H. Bandholtz who finally forced the surrender of Ola. But not until he had employed twelve companies of Scout soldiers and an equal number of Constabulary. The minor skirmishes against Ola were innumerable, and they occupied a period of many months. By the middle of 1903, Ola, Toledo, and Roldan had reached the status of definite menace to the peace. Ola's depredations in Albay, with 1,500 men, had been estimated to have caused a damage of $6,000,000 to the hemp industry.

In February of that year Captain Linsforth of the Constabulary of Sorsogon had met Toledo in a bloody fight when he found the bandit entrenched with 150 men on the border of Albay Province. Linsforth had rushed the position and had been repulsed with the loss of half of his command and the wounding of most of the survivors. In this engagement, Sergeant Vidal of the Constabulary, who had been a Major of insurgents and a companion of Toledo in the early days, stood up during the fight and called out in Spanish, "Toledo, do you know me? Come down with your *matcheteros* yourself." His challenge was not accepted by Toledo.

Roldan was killed by his own men during an attack by Lieutenant Cheatham after being badly routed by the Manila Company under Crockett. On September 25, Ola surrendered to Bandholtz and quite cheerfully turned state's evidence to hang most of his old associates. The following October, "Colonel" Lazara Toledo came in voluntarily and three of the most formidable gang leaders were removed from circulation.

When Ola surrendered to Bandholtz, he brought into the American camp an electric light bulb which was his personal *anting-anting*. He exhibited it proudly to Bandholtz, explaining its virtues as follows: "It has always been a sure warning of the presence of American troops near by. When I grasp it in my hand and the wires tremble, I know that the Americans are very near."

Bandholtz offered the suggestion that the hand trembled to shake the wires *because* the Americans were near.

H. H. Bandholtz was a great policeman and a magnificent soldier. He knew the natives as few white men were privileged to know them. He was a good friend of Manuel Quezon, and had he lived to see this Filipino lawyer become President of the Philippine Commonwealth, he would undoubtedly have held a position of good authority in the new Insular government.

Bandholtz was a man of great simplicity and he had a singular fixity of purpose. He wanted to be a soldier; so he went to a pretty good school, West Point, and learned how thoroughly. He had two weaknesses—candy and postage stamps. There was always a jar of candy on his desk and he was a stamp collector of note. He did not drink or smoke, and he was believed by many to be silent and unapproachable. Actually, he bubbled with a contagious enthusiasm.

He had a fine quality of winning the loyalty of his subordinates. "If you shoot that man, the bullet has to go through me first," was their feeling.

One of his officers writes of him: "Long after Constabulary days, I hunted up his outfit at Eagle Pass, on the Mexican border. He was in San Antonio, but I talked to his adjutant, whose face was glowing as he recounted the doings of Bandholtz."

There was great dignity in the appearance of this second Chief of Constabulary; he was of medium height, thick-set, with reddish-brown hair and piercing black eyes. He was a man to whom courage was a matter of course. He was the opposite of Garwood, to whom courage was a careless, romantic gesture.

Captain Higgins, who was adjutant at Lucena, remembers Bandholtz returning from a long leave in the United States. "He was as fat as butter; cheeks round and pink, and legs so plump that they bulged over his leggings. He was ready to start his Albay campaigns after Ola, following the failure of Garwood and Baker to apprehend the bandit. He cleaned up Albay. Six months later, he sent for me. All his fat was gone—he was as skinny as a rail and as brown as an Indian. He had lived in the saddle for half a year but Albay was clean."

As those Ola campaigns had approached the end, Bandholtz had gone into a town alone to accept the surrender of Ola and 1,500 men. He slept there, with his prisoners. We have a view then of a new and concealed joviality in his nature.

He came back with his prisoners—a long file of them—

with Bandholtz and Ola riding on the front seat of an escort wagon. Judge Carson, of the Court of First Instance, had been sent to Albay to try the prisoners. With Captain Higgins, Carson met the escort wagon and Bandholtz hailed him joyfully. "I want to introduce you to a friend of mine, Judge," he shouted. "Let me present you to Simeon Ola."

The judge came forward with hand outstretched, but when he heard the name Ola he quickly drew it away.

"Bandholtz," he said later, "that was a terrible thing to do. I can't shake hands with a man today and hang him tomorrow."

But Ola didn't hang—he was too smart to pay any penalty for a lifetime of brigandage.

Ola had brought in about 600 men, as the milder spirits had been turned loose immediately. Several hundred of these were in turn released. A few were tried under the vagrancy law and given road-work sentences of six months to two years. About sixty were sentenced to Bilibid for sedition, and twelve were hanged.

When Ola turned state's evidence he was given a 30-year suspended sentence.

In Cebu the Constabulary suffered a disaster. On a morning in May, 1903, Lieutenant Walker of the Constabulary of Cebu fell in his slender force of patrolmen and issued orders incidental to an extended jungle patrol. The inspection of combat packs, ammunition belts, and equipment was detailed, for Walker was carrying his small force into the very center of the domain of Quentin Tabal, who was bandit ruler of the island of Cebu.

After three days of hard marching the police detail was scrambling over the rocks of the interior, fighting jungle and insects and clammy waves of heat. They were nearing the center of the bandit country, and as yet had seen no hostiles.

Walker divided his force, placing six men under the command of Lieutenant Tuthill. The rest of the men he retained with him, and in two widely separated parallel columns the Constabulary began the ascent of a rocky hogback that led to disaster.

Halfway up the slope, the detachment of Walker came under a heavy fire from concealed riflemen. Bullets bit against the rocks and screamed across the valley. There was no target for the rifles of the police—the bush was without movement and the smokeless rifles of the bandits effectually shielded the location of the snipers.

Walker went down, drilled through the body. His men were like ants on the rocky slope as seen through the frenzied eyes of Lieutenant Tuthill, who tried desperately to break through the wall of riflemen that separated him from Walker's command. Across the wide valley he watched Walker and his men die, powerless to bring his men in support.

By superhuman efforts Tuthill broke through to Barili with his six men—Walker and his command were sprawled on the rocks in the bright sunshine.

With the killing of Walker the band moved north to Guimpilican, and Captain Page of the Constabulary went in pursuit. He arrived too late to aid Lieutenant Poggi in their extermination.

There, on the Tuburan coast, Lieutenant Richard Poggi fought one of the greatest jungle dogfights of the campaigns. On the twenty-first of May, Poggi took the field with a force of twenty-three Constabulary, and eighty volunteers under a civilian named McBride.

His way had been difficult, and his orders positive but vague: "Find the band that killed Walker." Poggi set out into the mountains along a narrow, twisting trail that wound ever higher through the gloomy passes. He knew that this patrol might be in the field for a month; he conserved his men, taking frequent stops for rest periods. As he climbed higher into the hills he put out advance points, hoping for quick contact with the enemy. From time to time his scouts brought in friendly natives; from these, he traced the route of the band that was giving way before his advance.

Night overtook him on the trail. He camped in a dense forest at the head of a narrow valley. There he received definite information as to the location of the outlaws. A native came into the camp and told him that the ladrones had a permanent fort that was high on the peak of a ragged splinter of rock.

Poggi rested his men for an hour and took the trail again. Near dawn, he was told that he was within rifle shot of the fortress. He deployed his men through the high grass, and like Indians they began a slow approach on the citadel that was the object of attack.

Even so valiant a leader as Poggi must have had doubts when he first saw that formidable fortified post that must be assaulted. It was an eagle's aerie, with one strong position above another. From where Poggi lay in the grass, the ground sloped up at an angle approaching the vertical to a broad ledge of rock where the outlaws had their first position. The Lieutenant estimated that his men, by clawing for each projecting bit of rock and each shrub, might have a chance to reach that broad ledge under fire.

The ledge itself was but the first step to the fort proper. From that narrow foothold, the bandits had raised bamboo ladders to provide a means of ascent to the summit. Once these ladders were drawn up, no force of infantry without mountain guns could force that precipice.

The bandits would have the Constabulary under clear fire as the police worked up the original slope. If rifle fire failed to halt the advance, the outlaws had only quickly to ascend the ladders to the summit and from that position pour a direct fire upon the unprotected ledge beneath.

Poggi divided his force, leaving half of the men in the protection of the bush, with orders to pour rapid and continuous fire upon the outlaw position. His whistle sounded and the Lieutenant led the remainder of the men to the assault.

Up that steep slope he took his men, with bullets leaving gray streaks on the rocks as they came under the direct fire from above. Boulders loosened by the outlaws crashed among his men as they hung by elbows and fingers and tried to return the fire.

They gained the ledge, and the outlaws had waited too long before beginning the retreat to the higher level. The advance of Poggi had been so rapid that he was able to pursue the outlaws on their own ladders. The battle that was fought there, hand to hand, on the swaying ladders, on the edge of that sheer cliff, goes down in Constabulary annals as an epic of jungle warfare. The wounded fell with the dead into the dizzy depths of the ravine.

When it was finished, 100 outlaws were dead on the field, and forty more were bound with *bejuco* ropes to be carried back to the station.

Quentin Tabal, the outlaw genius of Cebu, had been a menace since the year 1903. Here is an incident of that year to recall his reign of terror.

On July 9, a small Constabulary cart had rolled slowly into Cebu, with a guard of six weary Constabulary privates and a young Lieutenant. Even in those days of poor communication bad news traveled rapidly—this time the news had been such as to send a Lieutenant and six men deep into the jungles of interior Cebu. There they had found the slashed and hacked bodies of four young American schoolteachers who had been done to death by roving ladrones.

Twenty expeditions had been conducted by the Constabulary before the bodies had been recovered from the ravine where the outlaws had thrown them.

The murder of the four schoolteachers was only one incident in a series of atrocities and ferocious ambush that had marked the year 1903 as the most bloody in Cebu's history and brought into prominence Quentin Tabal, the outlaw. For three years, he was able to evade all efforts of the patrols to apprehend him.

It was soon discovered by the authorities that the elimination of the ladrone bands in Cebu could not be accomplished by patrol in the mountains. The brigands were difficult to locate and they seldom stood to fight. They had the advantage of being able to swoop suddenly upon unprotected *barrios* to keep their food supplies constantly replenished.

To curb the activities of the bands, the Constabulary had established a post in the center of the mountain area in an effort to bring into control some 5,000 people who were subject to the rule of Quentin Tabal. This measure was not successful. Tabal divided his forces and filtered through the country on all sides of the Constabulary station. The civilian population reached a state of desperate terror under the bandit raids.

A reconcentration measure was then adopted. All of the people of the district were gathered into fourteen small towns, each within two to six hours of the Constabulary post. Each of the villages was surrounded with a high stockade in the manner of frontier days in the West, and each village was established at a central point to control the lands they cultivated.

Each small group was then organized with a volunteer force of from forty to fifty men, armed with spears and bolos, to assist the Constabulary as lookouts, auxiliaries, and *cargadores* (baggage-carriers). For this service they were well paid by the provincial government.

This highhanded procedure of uprooting people from their homes and establishing them in arbitrary communities may appear strange to the reader, but it was the only means of assuring the protection of an unarmed civilian population. It had been authorized by section 6 of Act 781 of the Philippine Commission, dated June 1, 1903, which read as follows:

> "In Provinces which are infested to such an extent with ladrones that lives and property of residents in the outlying barrios are rendered wholly insecure by continued predatory raids, and such outlying barrios thus furnish the outlaws or ladrones their sources of food supply and it is not possible with the available police force to provide protection to such barrios, it shall be within the power of the Governor-General, upon resolution of the Philippine Commission, to authorize the provincial government to order that the residents of such outlying barrios be temporarily brought within stated limits of the poblacion or larger barrios of the municipality, there to remain until the necessity for such order ceases to exist."

Here again, in this law which the Commission had retained up its sleeve until the crisis developed in Cebu, was the reappearance of the old political intrigue which had contributed so much to the misinformation of the American public. An Act of Congress of July 1, 1902, had provided the government of the Philippine Islands, at the discretion of the Governor-General, could proclaim martial law at any time. Martial law was badly needed in Cebu, but the provisions of the Reconcentration Act avoided the exercise of this authority and kept up an appearance of peace.

The Reconcentration Act and the adjustments Cebu experienced under its regulation were successful in the elimination of outlaw bands without resorting to the armed intervention of regulars of the United States Army. Quentin Tabal and his brother, Anatalio, surrendered on July 13, 1906, and Cebu Island attained a state of quietude and relief from banditry.

The reader will recall the career of Ruperto Rios, the *Tulisan* who had established himself as the "Son of God" in a court that had bristled with high-sounding military titles. One by one, the retinue of "Generals" had fallen into the hands of the Constabulary, to be executed or imprisoned or pardoned, as their individual offenses merited. The "Pope" himself had managed to escape from engagement after engagement in Tayabas. The Constabulary had whittled away at his band in a series of mountain battles that had lengthened into more than a year of incessant pursuit.

During the year 1903, Major Murphy had crippled the strength of Rios badly in battles in the foothills of Tayabas, and the "Pope" eventually found the territory too hot for occupation. With his *Tulisans,* he crossed the border into Laguna Province and there he was captured by the active Captain Grove after several minor engagements.

Rios had a black record of murder and his end was foreseen when he came to trial. In August, 1903, the "Pope of Tayabas" was hanged at Alimonan.

Across the straits, on Panay Island, the Province of Antique was in the grip of a bandit leader named Opong. After a long career of outlawry, Opong decided to accept amnesty, and he came in to surrender to the Provincial Governor early in 1904.

He was released on parole, but upon his return to the mountains it soon became evident that Opong did not intend to honor the parole. Colonel Wallace C. Taylor sent Lieutenant Barry into the mountains to investigate Opong and the considerable disturbances that centered in his territory.

Lieutenant Barry found Opong to be the center of a cattle-stealing ring, and the discovery precipitated a fierce battle between the Constabulary patrol and the outlaw band. Opong and young Barry were killed and the band was broken up. Barry had borne a splendid reputation and his loss was a severe one to a force that had too few competent junior officers.

But peace came to Antique with the destruction of Opong, and never again was there serious disturbance in this economically poor province of Panay Island.

On the island of Leyte "Papa" Faustino was still maintaining his pretensions of divine birth. His citadel near Ormoc was the scene of complicated religious rituals. Faustino's communications went forth to all of Leyte signed "Señor Jesus y Maria," and calling for united co-operation against the government of the United States.

On August 24, 1904, Captain Barrett led a detachment of Constabulary into the mountains near Ormoc and came upon a *pulajan* fortress of "Papa" Faustino's that was defended with brass cannon. In the assault that followed, Barrett was killed by the full charge from one of these cannon and the Constabulary was repulsed. In this action the Lieutenant who had accompanied Barrett abandoned the attack and fled the field. He was dismissed from the service for an action that was a reflection upon the Constabulary and a damage to its prestige.

On the sixth of October, 1903, Lieutenant Velasquez took 30 men into northern Luzon, to make a census of the Calinga and Ifugao tribesmen. Not being the smoothest of diplomats, Velasquez was soon involved in difficulties with the hot-blooded mountaineers. While holding a conference he was attacked by 500 spearmen. His two sentinels, posted at the edge of the bush, were killed in the first sudden rush. Velasquez himself killed six of the seven chiefs about him with his revolver and dispatched the seventh with a bolo. His men formed in a close formation, back to back, and killed 53 Calingas and wounded 30. Revolvers only were used, as the action was too close for rifles.

The Constabulary then began a retreat toward Minerao, but were surrounded again on October 7 and had another running fight which resulted in the deaths of 30 additional Calingas. On October 8 the detail, now sadly depleted, was attacked by 300 Calingas, but they broke through the cordon by setting up a heavy fire concentration and managed to reach a station on the tenth, after being without food for four days.

For this futile action, Lieutenant Velasquez had brought severely to his attention the fundamental fact that the Philippine Constabulary was a police force, whose duty was to preserve peace and not to provoke hostilities. He was dismissed from the service, charged with lack of tact and failure to accomplish his mission without friction.

The Velasquez affair was but the first of a series of regrettable incidents that were nevertheless, valuable in the end in that they paved the way for new legislation and a complete reorganization of the corps.

Among the dangerous malcontents wandering the Philippine Islands in 1903, was one Artemio Ricarte who had been an active and inflammatory insurrectionist from the beginning of American occupation. He had been caught in the round-up conducted by the army, and upon refusing to take the oath of allegiance to the United States, he had been deported to Guam.

From this mid-Pacific island he made his way, in one way or another, to Hong Kong, and eventually he reached the Philippines again in the latter months of 1903.

He was an oil-tongued rascal with great persuasive powers, and he soon was able to gather about him a collection of flaming zealots who were violently opposed to every American institution. Ricarte soon decided that his most favorable opportunity for striking a blow at American prestige was to foment a mutiny in the Constabulary. He was fully aware of the hesitancy with which the Philippine Commission had originally armed this force. He began to play upon members of the Constabulary with whom he came into contact, with insinuations that they were a badly treated and half-trusted corps.

His poisonous propaganda began to have an effect upon the garrison at Vigan, and before the authorities were aware of the existence of treasonable activity, twenty-three members of the Constabulary had declared for open mutiny.

The uprising was quelled without serious disturbance, and Ricarte was sentenced to Bilibid prison, but the mutiny had given full play again to the old doubts of the advisability of arming natives. It was these isolated occurrences that kept the Krag repeating rifle from the corps for so many months.

Near by, in Misamis, the delinquency of two white officers of the Constabulary contributed to the unrest and distrust of the native police. Lieutenants Johnson and Hermann were jointly in a bad spot at their isolated station. Neither was of the fiber for patrol work, and in September, 1903, matters became serious when they faced exposure.

Johnson was short several hundred pesos in his accounts, and Hermann was facing charges of misconduct while acting as Senior Inspector of Bohol, which had been brought against him by Captain Green. On the night of September 27 they collaborated in the act of robbing the station safe and hiring a launch, the *Victoria,* they put alongside the steamer *Irene* and forced the Master of that ship to supply them with coal.

They then put to sea, carrying with them two enlisted men, who believed their officers to be taking them on a legitimate patrol. After a long sail to Bayauan, on the south coast of Negros, Hermann and Johnson went ashore to collect supplies. The unwilling Captain of the *Victoria,* assured now that something was wrong, sank his vessel on a sand bar.

When the Lieutenants returned, he reported that the vessel had sprung a leak and was unfit for sea service. At this point, one of the enlisted men made a great decision. He deserted his officers and hastened to report to the nearest Constabulary station.

Meanwhile, Johnson and Hermann and the remaining soldier had commandeered a native sailing ship and set sail with the intention of making the coast of North Borneo. Altercations sprang up, and in the night the crew, after being roughly handled by the Constabulary officers, attacked with bolos. Johnson was mortally wounded and the enlisted man was badly cut about the body. Hermann killed four of the crew with his revolver, and abandoning the vessel, took to the land, where he buried his stolen money.

Johnson, who soon died, and the wounded enlisted man were left in the deserted *prao.* For several days Hermann wandered in the jungle, trying to escape the rapidly closing cordon of Constabulary. He was captured on October 14 by a detachment under Captain Haskell, and was sentenced to seventeen years at Bilibid. Haskell also recovered the 1,800 pesos that had been cached on the lonely island.

These cumulative abuses brought about a prompt reorganization within the Constabulary. At about this time, new and very efficient men came into the corps. Captain J. G. Harbord of the regular Cavalry had been appointed an Assistant Chief on August 21. Captain William C. Rivers, also of the Cavalry, had been appointed as Headquarters Inspector of the corps.

With Rivers as Inspector, a new efficiency and a flavor of military discipline came into the Constabulary. Although River's health suffered in the tropics, he was a man of boundless energy. He was quick to detect dangers of isolated patrol service that did not come from contests at arms. Under his direction, the Constabulary gave up many of its casual ways. Rivers was a keen man. His was a great part in the restoration of the morale of men who had looked too long at jungle. The Constabulary became more military; the men had more duty as they rested in their jungle stations between campaigns.

By June 16, 1904, Rivers was ready with his General Order 75:

1. Practical instruction. There will be at each station at least two drills daily, except Saturday, Sunday, and holidays; a regular drill of one hour and, at a different time of day, setting-up exercises of twenty minutes. The regular drill will include the schooling of the Constabulary in infantry drill regulations and will be, when practicable, attended by all officers.

Rivers then devoted some space to the stipulation that officers should be careful in dress, appearing with belt and revolver outside coat.

His instructions continue:

2. When field work or patrol duties are not sufficient to keep the men in first-class marching condition, the drill will be lengthened to four hours once each week and devoted to a practice march of ten miles.

Inspector Rivers then mentioned that there would be a regular guard-mounting daily, and an inspection of clothing and arms each Saturday morning. He provided for instructions in first aid and for regular reveille and retreat formations. There would be no more idle hours of staring at jungle. A busy soldier was a happy soldier, as Rivers well knew.

The mental development of the officers and men came under the scrutiny of Rivers during this period. He improved the school of theoretical instruction for the men with subjects in Constabulary drill regulations, Constabulary orders, guard manual, military courtesies, method of securing warrants and making arrests, responsibilities of noncommissioned officers, geography of the Philippine Islands, and instruction in the English language.

Included in the mental examination was a recommendation from Rivers that all officers be paid an extra 100 pesos per year if and when they qualified in the Spanish language or in any one of the five main native dialects. Constabulary officers thus had an incentive to move down close to the people and learn their ways.

The officers were re-graded with a system of examination that improved upon the original schedules evolved by Captain Higgins.

The influence of Rivers upon the Constabulary can scarcely be overestimated. He came at a time when a man of his caliber was badly needed. Tall, lean, and intellectual in appearance, he was the scholar of the corps. A man of many words and of great friendliness, he was liked by his under officers, from whom he had the faculty of securing complete and willing co-operation. He had the screen doors removed from his office when he became chief of the Constabulary so that anyone who wished could enter and make known his desires direct to the chief.

As the author of the document that came to be known as the *Constabulary Manual,* Rivers laid the keel or backbone of the corps. His manual was the first work to put into handy form the manner in which the jungle police were to go about their duties and conduct themselves before a civilian population. It set forth in detail the laws and duties of a constabularyman. Many of the judges and officials of the period gave the *Manual* great praise and quoted it in their decisions.

The *Constabulary Manual* was brought out in Spanish and English, and it has gone through many editions since that day when Rivers made it the guiding law of the corps.

In 1906, Rivers was made a Colonel and Assistant Chief, to be advanced later to Brigadier-General and Chief of the Constabulary. Today he is retired in New York as a Major-General of the regular army.

The Constabulary of this period was not only attaining a state of high efficiency, but numerically, too, it was at its point of greatest strength. By Act of the Philippine Commission the strength of the force was increased to an authorized 7,000 to make provision for the opening of the Moro Province. Practically all of the Scout companies that had been detailed for service with the Constabulary had been turned back to the regular army. At the close of 1904 the

H. H. Bandholtz

Guillermo awaits execution

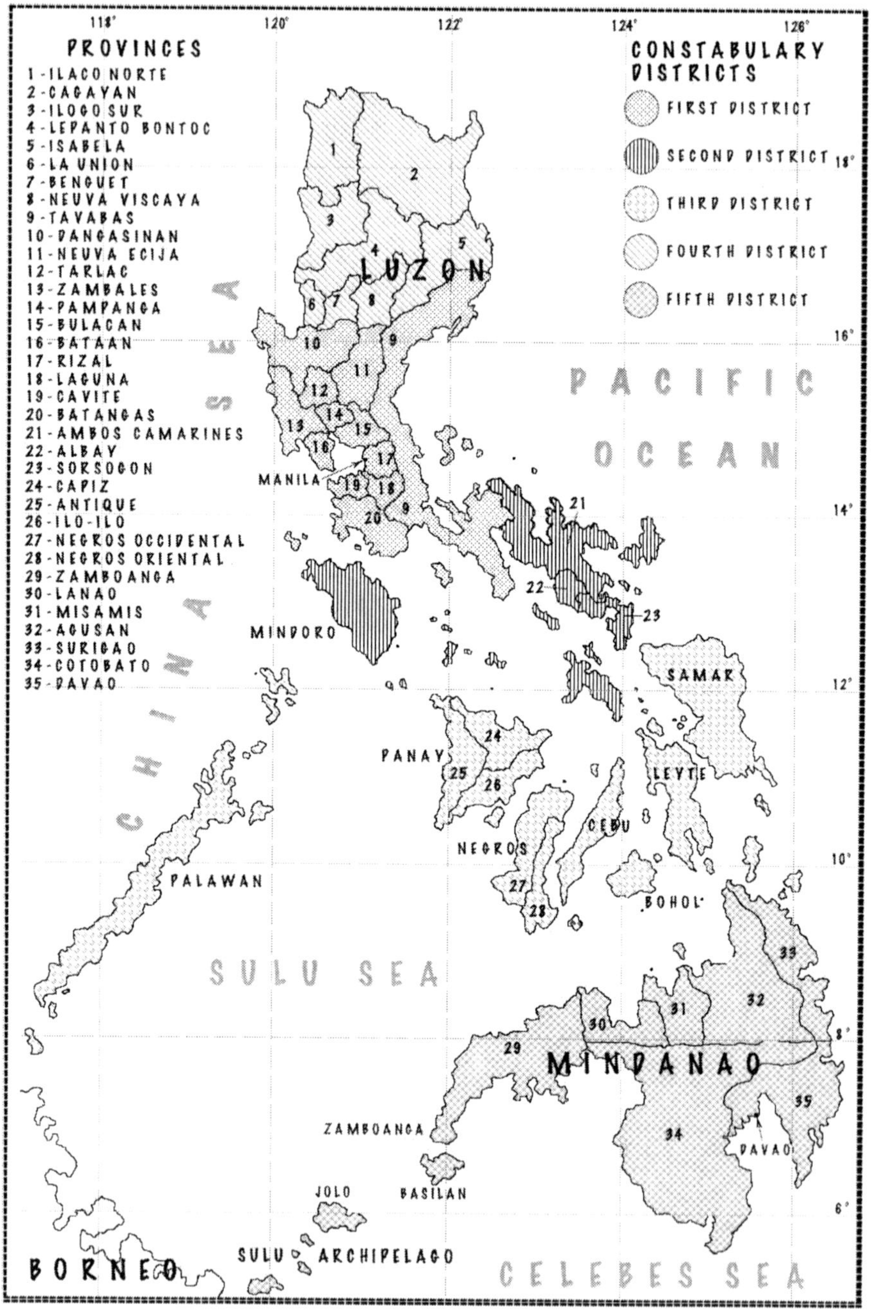

Map Redrawn by Christopher L. Harris based on Original Period Map by Paul H. Kirk

Philippine Archipelago

Insular Police had a strength of 288 officers and 6,950 men. They were occupying 250 posts in the archipelago, as against an army strength of 12,000 men in 39 posts.

There were in the Philippines, in addition to the regulars and Constabulary, fifty companies of Scouts totaling 5,000 men.

From time to time the Constabulary districts were changed, often so frequently that no attempt will be made to follow the constant geographic adjustments. In 1904, the districts were five in number.

First District

Headquarters at Manila. Colonel W. S. Scott, Assistant Chief in charge of the district. He had at his command 2,200 Constabulary on station in 80 posts, and an additional force of 1,700 Philippine Scouts who were not under his direct orders. First Lieutenant J. B. Clausen was Adjutant, and Second Lieutenant Thomas Leonard was attached to district headquarters.

Second District

Headquarters at Lucena. Colonel H. H. Bandholtz, Assistant Chief in charge of district. He had at his command 1,000 Constabulary on station at 43 posts, and an additional force of 1,200 Philippine Scouts who were not at his direct orders. Captain J. M. Wheat and Lieutenants C. D. Boone and C. C. Foote were assigned to district headquarters.

Third District

Headquarters at Ilo-Ilo. Colonel W. C. Taylor, Assistant Chief in charge of district. He had at his command 1,750 Constabulary on station at 80 posts, and an additional force of 2,000 Philippine Scouts who were not under his direct orders. Captains R. Vandam and T. A. Campbell and Lieutenants E. Walter, R. F. Adams, and W. C. Williams were assigned to headquarters.

Fourth District

Headquarters at Vigan. Major Jesse S. Garwood, Assistant Chief in charge of district. He had at his command 900 Constabulary on station at 33 posts. Captain T. C. Walker was assigned to district headquarters.

Fifth District

Headquarters at Zamboanga. Colonel J. G. Harbord, Assistant Chief in charge of the district. He had at his command only a few Constabulary, as the district was in process of organization and recruiting. The authorized strength of the district was 800 men, of which 530 men had been recruited in 1904 and were on station at 14 posts. Lieutenants J. Johnson and J. P. Caswell were assigned to district headquarters.

The average strength of a company of Philippine Constabulary was 2 officers, 1 first sergeant, 3 sergeants, 4 corporals, 4 first-class privates and 38 second-class privates. This strength was modified to suit existing conditions in the locality in which the company was stationed. It was seldom, in normal patrol, that a company of full organized strength was in the field.

By 1904, the arms situation had improved greatly. The ordnance records show on hand 7,370 Springfield rifles, single-shot; 2,251 Remington rifles, also single-shot; 4,072 shotguns, mostly single-shot, but included were a few repeating pump-guns; 5,129 Colt .45-caliber revolvers, double-action; 2,094 Colt .45-caliber revolvers, single-action, and 600 Colt .41-caliber revolvers.

On paper, this appears to be a great many firearms for a force of 7,000 men. Actually, there was still a shortage of arms, as the Constabulary was responsible for the arming of some 10,000 municipal police on station within the city limits of villages. These city police forces were very inefficient and required constant supervision.

The greatest difficulty was the securing of sufficient quantities of ammunition to allow target practice for the men. During the pursuit of Guillermo, the Constabulary had displayed an inability to hit their swirling, bush-screened targets, due to lack of target practice with their weapons. The police losses in this campaign were in excess of the quarry they pursued.

Until January 1, 1904, the rationing of the Constabulary had been an irregular and wholly casual arrangement. Sometimes rations were issued in small amounts at the beginning of a patrol, with the officers supplementing by purchasing supplies in the country using a *vale,* or I. O. U. system of bartering with the natives. This system was most unsatisfactory, as the natives disliked accepting the *vales,* often hiding all food supplies upon the approach of a patrol. By an Act of the Philippine Commission of January 1, 1904, the pay of all enlisted men was reduced seven pesos monthly and a ration allowance of twenty-one centavos per day was furnished. Thus each man was given a money payment for food and was expected to pay cash for his rations.

The system was so successful that it was later adopted by the regular army. By using this ration system the small patrols were able to penetrate far into enemy country without the necessity of carrying large food supplies. Ordinarily, when in settled country, the lure of actual cash was sufficient to cause food supplies to be offered gladly.

The bandit situation had improved in 1904 with elimination of Toledo, Roldan, and Ola, but there still remained the survivors of the gangs who reorganized under new, if less capable, leaders.

In the Cagayan Valley the Tomines rebellion burst to disrupt the public peace. Manuel Tomines was an ex-officer of the insurgent army and now was leader of a band that was well armed with forty Mauser and Krag-Jorgensen rifles. On the night of January 2, 1905, Tomines came down from the hills to make an attack upon the village of Naguilian, in Isabela Province. The force numbered 200 men, and the object of the attack was to kill one Señor Vicente Cauilan, a fearless Justice of the Peace who had prosecuted several members of the Tomines gang.

In the town of Naguilian was a small detachment of ten Constabulary soldiers under the command of a Corporal. Against this force Tomines took his 200 men. In a brief fight the Constabulary lost one man killed, and were driven across the river. Tomines and his men then turned to the residence of the Justice of the Peace. Finding the official absent, they ruthlessly butchered his wife, riddled the house with bullets, beat one of his brothers to the point of death, and carried away two women of the household.

The Constabulary immediately took the field, but it was not until March 27 that Tomines was captured as he attempted to enter the town of Naguilian for supplies. Notice of his arrival in the town was received by the Constabulary late in the evening, and a detachment immediately surrounded the house where he was hidden. Lieutenant

McLean entered by the back door and Lieutenant Collins by the front, while a Sergeant and two men remained outside to prevent the escape of the outlaw.

Tomines was found seated on the edge of a bed, revolver in hand but apparently dozing. McLean leaped across the room and wrenched the revolver from the bandit's hand before the weapon could be fired. In this coup Tomines and all of his gang but five were apprehended, and all of the participants of the bloody Naguilian raid were sentenced to death.

Tomines had as second in command an American deserter from the 16th United States Infantry, one Maurice Sibley. Following the capture of Tomines, Sibley took to the hills again, closely pursued by Constabulary patrols. Months later he was captured, to die on the scaffold.

In Surigao, the bandit Concepcion, who had raided San Jose, was hanged with four of his men, to bring comparative relief to that province. In Pangasinan, Pedro C. Gasig was coming into prominence. He had set himself up as a "Captain-General" with a force of ragged, homicidal followers. His reign of terror was short, for Lieutenant Haltman of the 34th Scout Company crippled him severely on July 19, and he was captured on November 11 by Lieutenants Basa and Gilheuser.

On the island of Samar, Pedro de la Cruz, who will be remembered as the *Dios-Dios* leader who was *Jefe Superior de Operacions* of all *pulajan* forces on the island, was stirring into activity. Now, in 1904, he fought a fierce battle with Lieutenant Clearman of the 39th Scout Company in which more than 100 men lost their lives. A few months later, Lieutenant McCrea took four men against de la Cruz in a gallant, completely hopeless charge that resulted in the deaths of the entire Constabulary detail.

De la Cruz disappeared then into the hills, to reappear no

more until he began to write his name in red in Samar two years later.

On Palawan Island, Lieutenant Louden of the Scouts suffered a severe loss when ambushed by hostile Moros while on a mapping expedition. Thirty-one carbines were taken by the Moros from the bloody field.

The close of the year found Montalon and Felizardo still at large; de la Cruz safe in the mountains of Samar, and the formidable religious movement in the middle islands growing in strength. In Mindanao, almost unpatrolled by the Constabulary who for some years were to play a very minor rôle to the army, the *Datu* Ali, scion of the ancient Mohammedan house of Kudarangan, was building a tremendous coalition of the Moros.

In Cagayan, during this bandit period, occurred one of the most grisly incidents of the entire American occupation of the Islands. One man, by name Daniel Verzon, had gathered three kindred spirits and begun a career of banditry. His course of crime was brief and unprofitable, for he had hardly more than taken the outlaw trail before his band was captured by the Constabulary. The four men, Daniel Verzon, Nicholas Ancheta, Faustino Pascual, and Anacito Javier were brought into town for execution.

For some reason which remains unexplained, the method of execution chosen was the garrote. The reader may or may not be familiar with this ancient Spanish method of punishment. The subject is seated in a chair, his back to a strong post to which is affixed an iron collar. This collar encircles the neck of the victim and it is slowly tightened, under pressure of a screw that forces itself into the nape of the neck and accomplishes the twin results of strangulation of the victim and dislocation of the spinal column. It is a slow, agonizing death, in the best Spanish manner.

BANDITRY

There was in this little town of Cagayan, an ancient garrote in the town square that was a relic of the Spanish occupation. The four unfortunates were led to it for execution. Now the garroting of men is not an American institution, and the executioner who had been imported from Manila proved entirely unfamiliar with the device of execution. Each victim, in turn, was seated in the chair and after eight minutes each of manipulation with the screw, the bandits were pronounced dead.

As was the custom of those days, the bodies were laid in the square as an example to other evildoers. Some three hours after the execution a casual passer-by saw a most horrible sight. Verzon, Ancheta, Pascual, and Javier were twitching feebly in great agony; they had survived the fumbling efforts of the amateur garroter.

Everything was done for them then by the doctors. They were removed to the homes of friends, and although they lived, they were totally paralyzed and pronounced incurable. So they remained for a period of many months following their "execution."

Legally, they did not exist. Garroted on October 30, 1903, they were still alive on June 30, 1904, when the official report of their execution appeared in the annual summary of the Chief of Constabulary.

As one reads the combat reports of this bloody era, it would seem to be mostly a chronicle of death. Death on the blade of a bolo; death from cholera and dysentery and drowning; death from Springfield and spear and hangman's noose. Human life was of such small account in those days of jungle conquest.

There were abuses, not only in the conduct of the Constabulary but in the record of all of the units of armed force in the Islands. In the bush, all Filipinos looked alike. The innocent went down sometimes with the guilty, during that day of hand-to-hand battle that offered no quarter. There are always abuses in the conquest of a raw, wild land. The unwilling colonization of a wild people is in itself an abuse.

But striking averages, one is impressed, in the main, by the conduct of the Constabulary in these difficult days. They were trying to do a job to which they had been ordered. Sometimes there was an element of chill ferocity in the manner in which they conducted their jungle campaigns, but it was sub-surface warfare they waged. They killed to live. It was treacherous warfare and shielded by jungle and only the most hardy could survive those battering years that the Constabulary spent before the open blades of fanatical natives. It was certainly not an age for sentiment.

But we look in vain for evidences of the slaughter of natives by an overwhelming force of Constabulary. Always the ratio was in reverse. True, the Constabulary participated, with a few men, in the Moro massacre at Bud Dajo, but theirs was a minor part in that battle. Apparently the slaughter the Constabulary performed was a desperate, defensive slaughter against men in vastly superior numbers who came on and on until they were dropped by rifle fire.

But even in that grim period when raid and killing was the rule, there appears occasionally a story of the lighter side of the jungle soldiers.

There was a young Lieutenant who had captured a small detachment of Filipinos with arms, ammunition, and a trumpet. With conscientious pride he prepared a report. "Captured a detail of six Filipinos with their arms, ammunition, certain supplies, and a trumpet."

A helpful, alcoholic comrade found the report on the Lieutenant's desk, ready for mailing to headquarters.

Thoughtfully, the inebriate added one letter to the report and forwarded it to Manila.

A week later, the officer who had made the capture had the report returned to him by a genial commanding officer. It was blue-penciled with the remark, "How did you know that she was a strumpet?"

At the close of that bruising bandit year, the administration was able to report: "The Constabulary are to be congratulated upon having so far reduced the ladrones that it is now safe to travel practically throughout the archipelago."

In this optimistic report, the Commission was overenthusiastic. The outlaw bands, in truth, had been badly scattered, and many of the leaders had been executed. But with the passing of the ladrones, a new and far more dangerous menace had come into being. The middle islands were swarming, as this Commission report had been written, with religious bandits whose activities were to surpass the best efforts of insurgents and ladrones. The Constabulary was entering its third battle phase—the conflict with the "Popes."

And in the south were the Moros!

Chapter viii

PATROL

> "On marches and expeditions officers will march with their commands; they will not be mounted unless the men are . . ."
>
> —*Constabulary Manual*

DURING 1905 and a portion of 1906, the Constabulary was engaged in the elimination of the last of the leaders who might be charitably considered as "insurgents," although their activities savored more strongly of banditry and murder.

Some of these men had been dangerous figures since the days of Spanish occupation. Their organized depredations had grown to such proportions by 1905 that it became necessary to call in the army, and to set up a provisional district of Constabulary to suppress them.

It will be remembered that the notorious Felizardo and his confederate, Montalon, were still at large in the north in co-operation with other leaders of the old "*Partido Nacionalista*" movement. They now burst upon the scene with a final sporadic series of grim and murderous raids.

In December, 1904, matters had become intolerable within a few miles of Manila. With armed bands roving within seven miles of Malacanan Palace, where the Civil Government sat in solemn session, Governor Taft at last bowed to the inevitable and declared a state of martial law, with regulars in the field.

A large band of armed bandits descended upon Paranaque, seven miles from Manila city limits, and raided the Constabulary station of that town. The victory was complete. With arms and ammunition taken in that raid, the bandits then made their way to Taal, where they looted the municipal treasury of 15,000 pesos and carried away all the weapons of the police force.

These same ladrones, then, uniformed in captured tunics of Scouts and Constabulary, raided San Francisco de Malabon in Cavite Province. Here, on January 21, they rushed the *cuartel* and secured twenty-one rifles and a great store of ammunition. Dr. J. O. Neill, medical officer of the Scouts, was killed by rifle fire while escorting his wife and daughter to safety. The outlaws tried to capture Mariano Trias, who was the Provincial Governor. Failing, they carried away his wife and daughter, who were later released when Constabulary pursuit became onerous.

The prestige of the bandits had grown to such proportions that on January 31, 1905, it was decided to set up a Provisional District of Constabulary within the provinces of Cavite, Batangas, Rizal, and Laguna, with a special force under separate command.

The area of greatest unrest centered about Cavite, the Tagalog province across the bay from the city of Manila. The region is a rolling country with few mountain areas, sparsely scattered with jungle, and covered for the great part with fields of high *cogon* grass. For four centuries the province had been overrun by cattle thieves and marauding bands, with the Spaniards unable to remedy the situation. From Cavite Province the outlaws had slipped over to Batangas, Rizal, and Laguna, and a reign of terror held the native population in a state of virtual servitude to the fences of stolen merchandise who were called *pillos*. Quite often these *pillos*, who acted as go-betweens, did not know the real leaders of the outlaw bands they served.

In these four provinces had grown to strength divers outlaw bands, all pseudo-political, and all owing allegiance to an ex-barber named Marcario Sakay, who was a self-appointed "President of the Philippine Republic."

Although a military organization of bandits was ostensibly present, with a sworn purpose to facilitate Philippine independence, the gangs were actually merely brigands. They wore no uniforms, and when forced to cover by the authorities, were accustomed to hide their arms and melt into the civil population.

It was to combat the bandit leaders, Montalon and Felizardo, underlings of Macario Sakay, that the Constabulary took the field.

In December, 1904, it was demonstrated that the slender police force could not cope with the emergency. The wave of banditry was beyond the capabilities of the Constabulary. After several months of inactivity, the regulars were called into the field.

The Constabulary were placed in command of Colonel D. J. Baker, Jr., and the writ of habeas corpus was suspended. General Corbin of the Philippine Division sent the third squadron of the 2nd Cavalry and the 4th company of the Infantry under Major F. W. Sibley to assist the Constabulary in restoring order.

With this regular army force augmented by a strength of 1200 Constabulary, a concerted action was undertaken. The oulaw bands were several in number. "General" Oruga, who still maintained pretensions of bona fide insurgency, was in command of the forces of "Colonel" Villanueva, "Lieutenant-Colonel" Vito, and "Major" Flores. "Major-General" Felizardo had the bands of "Lieutenant-Colonel" Caro, "Lieutenant-Colonel" de Vega, and "Major" Giron.

This entire Filipino force was consolidated under the supreme command of Montalon as "Captain-General of the Armies of Liberation."

In a series of combined Scout-army-Constabulary operations, Oruga was crushed by Thompson and Baker; the United States cavalry detachment killed Caro in a fierce night fight at Tres Cruces, and Van Schaick and his Scouts pounced upon Felizardo at Lit-Lit. From this net Felizardo escaped, to be engaged again on June 14, 1905, by Lieutenant Lorenzo Ramos of the Constabulary. Ramos captured the outlaw camp and killed several of the outlaws, but Felizardo melted as usual into the bush.

Captain Van Schaik of the Scouts continued the pursuit with a night attack on June 21. The morning of June 22 found Felizardo wounded and half naked in the wilds of Cavite near Bacoor, but two men with him.

For sometime the trail was lost, but the Constabulary never relinquished its efforts, and one day in the little *barrio* town of Batangas, a bandit appeared and demanded food from an old Filipino. The Filipino tapped him on the head with his rice pestle, killing him. He and his companions took the body to a cliff and threw it over. They hiked into the nearest Constabulary station and reported that they had surrounded a bandit who had leaped to his death rather than be taken. Baker was so impressed with the bravery of a ladrone who would jump to his death rather than be taken that he ordered the body sent through to Tanauan for indentification. Many natives looked at it and said, "Felizardo." The last man to identify it said that he had a broken tooth. Baker was convinced that Felizardo was dead, because they opened the bandit's mouth and found the broken tooth on the left side. The 5,000-peso standing reward was paid.

And as a conclusion to the campaign, Van Schaik was sent to the Province of Cavite as Civil Governor. He had not been there very long before he crossed to Manila to have lunch with his old friend, Harry Belden. Mrs. Belden said, "My washerwoman told me a remarkable story. She said she was in Bacoor the day Felizardo's funeral took place. 'I saw Felizardo's mother,' said the washerwoman, 'and as the body came down the street I stepped out and put my hands around the weeping mother's shoulders. "Do not cry anymore. Nothing could be done." The old mother, without turning her head spoke out of the corner of her mouth and said, "Get out. I saw him half an hour ago." Then she let out a fresh wail and followed the body down the street.' "

The story had a ring of truth, so Van Schaick left at once for Constabulary headquarters to check on the washerwoman's story. At headquarters it was decided to start all over again and to question his death until the definite indentification of his body.

A fine officer named Aurelio Ramos and two enlisted men, started the campaign. Ramos abused the two soldiers at drill in the presence of witnesses. He let them work all day in the hot sun on the parade ground, and the two men would at night go down to a *tienda* where an old mistress of Felizardo lived. Ramos further set the trap by refusing to trust the rest of his men, taking all their guns away at night and locking the arms up in a little *nipa* house. The two soldiers went to the woman of Felizardo and told her that they were deserters and asked her where they could find Felizardo. She told them to go to Bacoor to the house of Captain Damaso Mareuz. They went, and Damaso told them to go to a certain point in the woods at three o'clock in the morning and camp until daylight. They followed his instructions, but nothing happened. They went back to Damaso. He told them to do the same thing the next night. Out of the shadows came a guide to lead them to Felizardo. They told Felizardo how badly they were treated, how crazy Captain Ramos was, how foolish he was to put all of the guns where they could be stolen so easily. Finally they suggested to Felizardo that they go to Camp Nicolas and capture the guns. Felizardo consented, and turned out fifty *barrio* people, who helped him. Felizardo told the *barrio* people to wait and took the lead himself, with the Constabulary soldiers following him closely. As they passed under the shade of a great mango tree, the two soldiers laid down their guns, because the last time Felizardo was killed, he did not stay dead and they wanted to be sure and finish him this time. After they dropped their guns, the two soldiers closed in on Felizardo. The arm of one of them went around his neck from behind. The fight was a terrific one. Finally one soldier had been kicked entirely to one side, and as the other started to come in again, Felizardo half lying, half sitting, stopped them with the very force of his personality. He said, "Carpio, you are my cousin. I did not expect this from you. Don't you think we had better stop and talk it over?" Carpio weakened, but the other man who had been kicked to one side reached for his bolo, and creeping up from behind, crushed Felizardo's skull. Then the two soldiers paraded to their camp and Ramos, with a group of men, surrounded the mango tree. He ordered them to step forward and pick up the body. Not a man moved. They still felt that Felizardo would not stay dead. Ramos himself lighted a flare, and took the body and shoved it so as to show the men that Felizardo was dead. Ramos and one of the men climbed into a little *carromata*. They put the body on the seat between them and drove twenty miles to Manila. There, the body was positively identified by a little thirteen-year-old boy named Lucino who had been Felizardo's *muchacho,* as well as by other people. So the second reward was paid to the men of the Constabulary who had rid the country of another notorious bandit.

There is every indication that Felizardo himself collected

the first reward paid for his death, obtaining it through an agent. The body of the man turned in for the reward is believed by many to have been killed by Felizardo himself.

Lieutenant Walker of the Scouts and Lieutenant McLean of the Constabulary were conspicuous in these northern campaigns. The fighting closed with the surrender of Montalon. With the "Captain-General" came his lesser satellites, Carreon, Benito Natividad, Lucio de Vega, and Leon Villafuerte. The most important capture of all had been the taking of Macario Sakay, who was the organizing genius and unofficial commander of the rebellion.

The next year the unsavory quartette were tried for murderous offenses that stretched across a decade. Montalon and Villafuerte were saved from execution, to be given a life prison sentence. Sakay and De Vega were hanged.

The extent of the reign of terror occasioned by these four men can be gauged by the testimony at the trial and the findings of the court. The transcript of the trial shows that the bands had "killed a constabulary private on a peaceable survey party; hung Lorenzo Amigo, a resident of Caloocan; brutally cut the tendons of the hands and feet of Natalio Anitares and Candido del Mundo and then slew them; slashed to death Tomas Panuelpa and his brother; shot Benigno Martin and Teniento Juan at Bacoor; hung Melicio Alcantara and Alejandro Jesus; poured petroleum over Patriarco N. and burned him to death; hamstrung and cut off the lips of Blas Cabrera; cut off the upper lip and severed the tendons of the right foot of Martin Piol; hamstrung Vicente Castillo and Isidro Camiuac; mutilated and crippled for life Simeon de Quiros and Calexto Rollo; hamstrung German Oliveros; captured two female servants in Malabon and repeatedly outraged them; carried off and outraged Rosa M. of Tanauan and while resisting her rescue killed policeman

Francisco Guevera and wounded Sergeant Gonzales; seized the father of Justa M. of Bacoor and under threats of death obliged him to withdraw his 13-year-old daughter from a convent in Manila and deliver her to the brutal embraces of Felizardo, and finally carried away the wife and two baby children of retired General Trias and in the depths of the mountains submitted this gentlewoman to treatment worthy only of brutes and savages."

Prior to and coincident with the Cavite and Batangas campaigns after Felizardo and Montalon, the Constabulary were elsewhere engaged in writing final chapters to the careers of many of the northern bandits. In Isabela Province, Manual Tomines was executed, and his aide, Maurice Sibley, the American deserter from the 16th infantry, was sentenced to a long term in Bilibid prison.

The outlaw Sarria, who had been affiliated with Ola and Toledo, was killed in Ambos Camarines by Captain R. H. Griffiths, and this same officer removed another menace to the peace in the person of Francisco Gamboa a few months later.

Meanwhile, Faustino renewed his depredations, and it was not until April, 1905, that Captain Grove, after an eight-day march, corralled Faustino and most of his band in a blockhouse. Eleven of the *pulajans* were picked off by sharpshooters and the expert pistol work of Grove, and the balance fled to the hills before the spirited charge of the Constabulary which resulted in the capture of the blockhouse.

Then on July 28, a Constabulary unit under Lieutenant O'Conner located Faustino in the center of the mountainous region between Dagami and Ormoc, where the "Pope" had built a strong fort defended by 100 men and several brass cannon. O'Conner attacked with a small force, and after a savage assault and siege that lasted for more than three hours, Faustino was wounded in the chest, his two sister "Saints" were captured, and thirty-two of his followers were slain.

The "Papa" escaped again, and after a long series of operations, with Lieutenant Jones of the 8th Infantry conducting a campaign and Lieutenant Snodgrass again wounding the "Papa," Faustino was at last brought to bay for the last time. He was killed, with twenty-one of his men.

The other serious menace to the peace of Leyte was a ladrone named Juan Tomayo, who operated in the lowlands of Jaro and Garigara with a small band of ruffians. His course of murder was cut short by the municipal police of Zumarrange, on the island of Buad, which is opposite Catbalogan Samar. He was killed there in a raid upon the municipality, by the efficient city police.

These campaigns against the ladrone chiefs were grim and unromantic affairs. During this entire period of guerrilla warfare there was hardly a day that the patrols remained in barracks. The splendid organization of some of the bandit gangs made them worthy foemen, and the records of the period are filled with examples of fanatical bravery on the part of the ladrone chiefs. The weird religious rites they practiced apparently had some effect upon the weaker members of the enlisted personnel of the Constabulary, for in Colonel Taylor's report of 1905, we find the following:

> "Following an engagement near Ormoc, two privates were heard to say, 'There must be something to the *anting-antings* as we fired repeatedly at *pulajans* and never hit any.' They were discharged for expressing this belief in the powers of the *pulajan anting-antings*."

Other than these minor delinquencies, the morale of the Constabulary was excellent, for there is record of but nine desertions in the period from June 15 to November 15, 1905.

The destruction of "Papa" Faustino and Juan Tomayo had been a long step in the establishment of order in Leyte, but the virulence of the *pulajan* movement was not expended with their elimination. New leaders arose to replace the old, and the bands were soon consolidated again to renew hostilities.

The next year the *pulajans* came down under the guidance of Felipe Ydos to assault the town of Burauen, Leyte. Their attack upon the police barracks resulted in the deaths of five Constabulary and the wounding of seven. The *pulajans* withdrew with fourteen rifles and a quantity of ammunition.

Lieutenant L. E. Jackson took twelve men in pursuit of Ydos, and Major Henry C. Neville was in the field with two companies. This combined force combed the jungle for a week with no trace of the raiders to be seen. Finally, on July 5, Major Neville, with forty men, attacked a strong *pulajan* position near Mantagara, west of Burauen, and lost seven men in pitfalls filled with erect spears. He was unable to capture the position and sat down to wait reinforcement.

On July 11 the *pulajans* abandoned their fort at the approach of a strong body of Constabulary, and a week later they appeared before Patock. Captain Beazeley was in pursuit, but after two days he lost the trail and retired from the field. Lieutenant Williams had a short fierce battle west of Burauen with indecisive results.

Then, late in July, the band was reported in the vicinity of Burauen, and Lieutenants Williams and Worswick went with thirty-four men to investigate. Two miles west of the

Burian-Dagama road they were attacked in force by a band who opened fire from ambush.

After a scattering volley from the jungle edge, the Constabulary was rushed on the left flank by bolomen. As they retired, twelve soldiers, Lieutenant Worswick, and the American civilian, McBride, were killed.

McBride will be remembered as the battling civilian who had accompanied Lieutenant Poggi in the assault of the *pulajan* fortifications on Cebu. Constabulary and army records frequently conclude with the sentence, "They were accompanied by a civilian named McBride."

The man was a true soldier of fortune—a strange figure who has marked the pages of Philippine history. He owed allegiance to no recognized group of fighting men—he bobs up here and there, apparently serving indiscriminately with Scouts or army or Constabulary. He was found where the action was heaviest, and he was famous for his stock of bush lore. In passing, one wonders about this unique, romantic figure who wandered about continuously seeking a fight.

When McBride went down before the flashing *pulajan* blades near Dagami, the surviving American, Lieutenant Williams, broke through to report the disaster.

On the following day, Neville and Lieutenant Jeancon, with fifty Constabulary, joined forces with Captain McMasters, who had a platoon of Company E of the 24th Infantry. With them, another platoon under Lieutenant Silcox arrived at Tabontabon to continue the pursuit.

At Tabontabon, the *pulajans* were located and immediately attacked by the American forces, who inflicted a crushing defeat on the fanatically brave *pulajans*. The actual count, after a limited search of the thick bush, was forty-nine natives killed and three wounded.

A few months previous to this action, Neville had walked

with two of his men into the camp of a band of fifty-seven armed outlaws. He was far in advance of the main body of his command, and finding himself in the camp of the men he was pursuing, he began to issue sharp orders as though he were in complete mastery of the situation, thus holding them for forty minutes until his support arrived on the scene. He was the finest pistol marksman of the corps; even the irrepressible Garwood bowed, in open competition, to this silent Major. Neville was noted for great coolness under fire and a methodical disposition of his men in positions to retain every possible advantage.

On June 23, 1905, Captain James R. Lewis was awarded the Medal of Valor. . . .

A small detail of the Constabulary under Lieutenant Harry L. Beazeley was in co-operative service with a volunteer force of bolomen under Captain Lewis. They were high in the mountains of the Leon country of Ilo-Ilo on a special mission that was concerned with the capture of the noted outlaw, Tomas.

Tomas had been an elusive shadow who struck and vanished, and returned unexpectedly to strike again. The Constabulary was eager to find Tomas. They found him, that morning in 1905.

As was their custom, the patrol was pushing into the jungle with flankers out in anticipation of ambush. The flankers had cut their way through the rank bush but they missed Tomas, who had been there, shielded by the liana-hung forest trees. The sound of shots . . . the thud of bolo blades; Tomas was upon the detail before a retaliatory shot could be fired. It was one of the few skirmishes of the period that was fought almost entirely with blade weapons.

The bandit hurled himself upon Captain Lewis as his

special object of attack. The forest was filled with sound as bandit and Constabulary soldier paired in individual combat.

The rifle of Captain Lewis jammed as he attempted a snap shot at the oncoming bandit chief. He hurled it from him and drew a blade from his belt to oppose the swirling bolo that reached for his head. Back and forth they twisted and dived and feinted and plunged—a strange duel to the death, with white officer and native bandit measuring their strength with the weapon of the country. Bolo against bolo in that blood-slippery jungle clearing.

With a supreme effort Lewis cornered the shifty Malay and almost decapitated Tomas with a blow. As Tomas fell, Lewis twisted in mid-air to face a new menace. On the ground, a wounded bandit was drawing a bead with a rifle on the Constabulary Captain. Lewis severed the head from the body before the bandit could pull the trigger.

In the few minutes that Lewis had fought for life, bullets had passed through his hat, grazing the skull: through his shirt, between the left arm and the body; and through his trousers, wounding the left knee.

With the destruction of Tomas and his band, a leader with a long record of murderous assault was eliminated, and a long existing menace to life and property was removed from Ilo-Ilo Province.

Cary Crockett

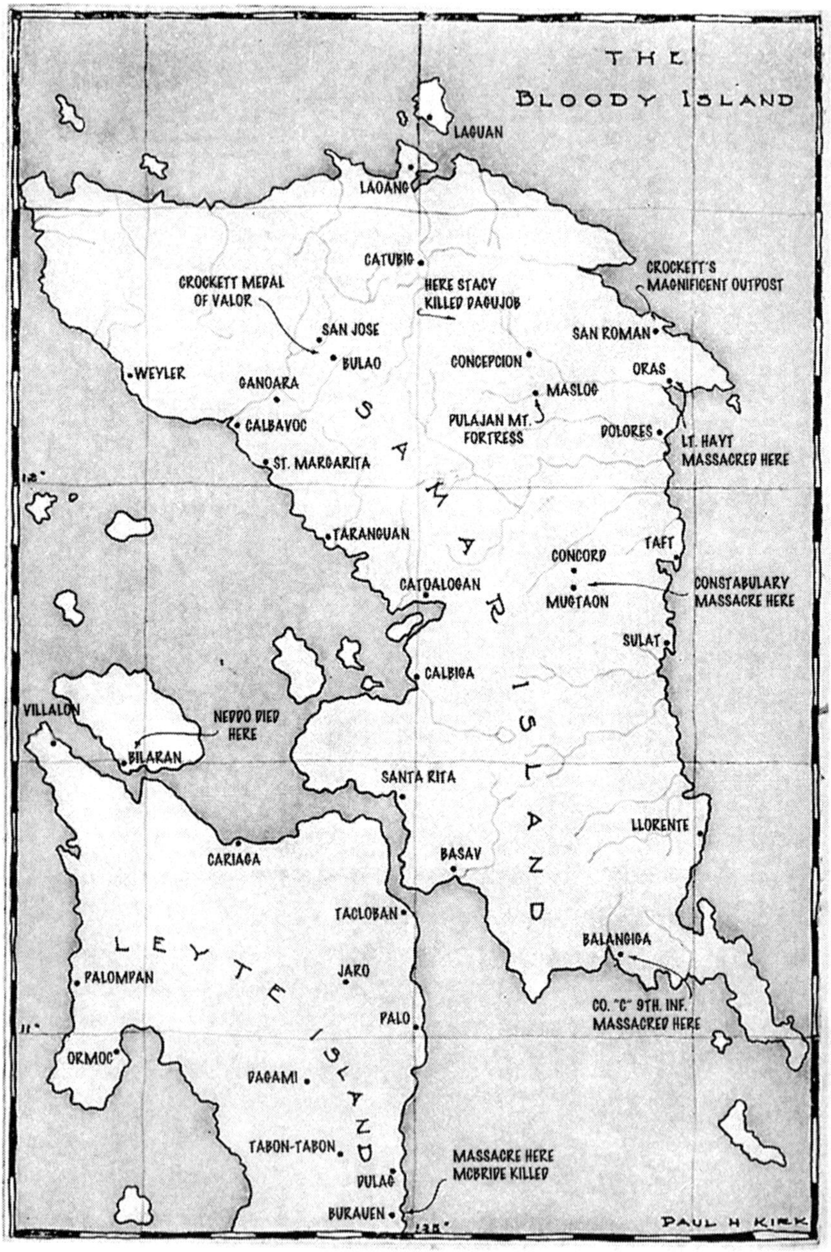

Samar Island

*Original Map drawn by Paul H. Kirk for 1938 Edition of **Jungle Patrol***
Enhanced by Christopher L. Harris

Chapter ix

THE BLOODY ISLAND

"Patrols should always sleep in a place from which they can inflict injury on a party attempting to surprise them . . ."

—*Constabulary Manual*

SAMAR was a place of evil reputation.

The army remembered it for the massacre of Company C of the 9th United States Infantry at Balangiga. The Scouts were to remember it for the slaughter of Lieutenant Hayt and his entire detachment of forty-seven men. The Constabulary would write Samar into their records as the arena that would send many men with red epaulets to their deaths. The army had written a song about Samar, to commemorate that dark and bloody island: "There's many a man been murdered in Samar."

We have seen the army preceding the Scouts and Constabulary into Samar, to conduct a series of very severe operations under the command of General Jacob Smith. This General had acquired an unfavorable notoriety because of his famous orders "to kill everyone capable of bearing arms and to leave Samar a howling wilderness." When he had been pressed for an additional interpretation of the order, the General had established the arms-bearing age as "everyone over the age of ten."

This campaign of the regular army had resulted in the court-martial of General Smith, but not before the regulars had finished a bloody operation that ranks as the most ferocious performance of the army in the Philippines. Samar *had* been left a howling wilderness, but not precisely in the manner that General Smith had intended. The army had eliminated certain bands of organized insurgents but the religious bandits were unscotched, and the hand of civilization rested lightly indeed upon the island of Samar.

If one turns to the map of the Philippine Archipelago, Samar can be seen there: a great thick bulk of jungle island almost touching the jutting peninsula of southern Luzon. The interior of the island is a place of high mountains, with sheer ravines and swift, rushing rivers. In the rainy season, the floods—the dreaded *avenidas*—pile the water high in the gorges and sweep everything from their path. Not an inconsiderable cause of army and Constabulary casualties was the loss, by drowning, of men who had been swept away in the *avenidas*.

Interior Samar was a place of great snakes and malaria mosquitos and sludgy, oozing swamps on the fringes of the forested mountains. Here and there were the abandoned *caingins*—the clearings of the mountain tribes, grown high with lush *cogon* grass and *tigbao*. It was a place calculated to try the stamina of the fighting men who forced the bush of the interior.

The island has an area in excess of 5,000 square miles, and at the time of which we write, there were not five miles of road on the island.

The army maintained several stations on Samar, but the troopers were inactive and confined to station by the orders of the administration, who were still trying to avoid the necessity for martial law and the massed movements of regular troops.

On July 1, 1904, the military forces of the United States in garrison on Samar consisted of one regimental post at Calbayog and a company of Scouts on the north coast at Laoang. The Constabulary were in the district with 32 men on the north coast at Catarman under Lieutenant Bowers; 78 men on the east coast at Borongan under Lieutenants Poggi and Abenis (Poggi was temporarily on station in Mindanao on detached service); 10 men on the west coast at Calbiga under Lieutenant Sulse; 80 men at Catbalogan under Captain Hunt, with Lieutenants Jeancon, Burbank, and Martin, and 39 men on the south coast at Balangiga under Lieutenants Smith and Farrow. In addition to this force, Company A of the Manila unit was on field service under the command of Captain Cary Crockett.

At this time (1904) the real force of the *pulajan* movement had not been realized by the administration in Manila. Although the jungle was swarming with these fanatics, the reports of their activity are very brief. An official report dismisses the *pulajan* trouble on Samar with this brief remark: "The *pulajan* band of Pedro de la Cruz, with nine rifles and usually about sixteen men, is the only band of importance on the island. A *pulajan* leader by the name of Anugar also circulates in the Gandara Valley." Actually, could the administration have known, there were on Samar several thousand armed *pulajans* who were consolidating and organizing deep in the mountains of the interior.

In the fall of 1902, after the surrender of the insurgents on the island to the military authorities, Anugar and other less important leaders had addressed communications to people formerly identified with the insurgent movement, calling upon them to come to the mountains and continue the fight against America. They stated that they had not surrendered to the United States and never would, and that the mountains of Samar held thousands who were ready to begin fighting at any time under the leadership of the "Popes."

As time went on, the influence of these agitators among the mountain people began to be noticeable, and isolated Constabulary officers on patrol duty began to make reports to the Philippine Commission which were for the great part ignored. As these reports filtered into headquarters, the only change instituted on the island was the establishment of another Constabulary post at Bulao, and a post of Scouts at Gandara.

In February, 1904, Pedro de la Cruz made an appearance near Borongan, and Lieutenant McCrea, while scouting with a patrol of seven men, was defeated and killed by the outlaw chief. A few days later, a detachment of Scouts recovered the bodies and fought an indecisive engagement with the same band. The rifles were not recovered.

This affair served to direct attention to the *pulajan* movement, but the impetus of the fanatical organization was still improperly gauged. De la Cruz had been apparently only trying his strength, for he retired again to the hills and was not heard from for some weeks. Small Constabulary detachments operated freely in the country without molestation.

The real trouble broke early in July, 1904, with an attack by the *pulajans* upon isolated *barrios* of the upper Gandara River country. Villages were burned and the civilian population subjected to horrible atrocities. A force of Constabulary was sent up the river to quell the disorder, and a grievous mistake was made by the officer in command in detaching twenty men, under a Sergeant, for station in the infected region.

This detachment was attacked and driven down the river with a loss of eight rifles.

The defeat of this Constabulary section was the flame that touched off the rebellion. The headwaters of the Gandara became the scene of very severe fighting, as detachment after detachment sought vainly to restore order.

On August 15, 1904, General Allen ordered Captain Cary Crockett, with a company, to take station at San Pelayo and to conduct operations against the *pulajans* in the upper reaches of the Gandara.

Meanwhile, the *pulajans* had burned the town of Bulao.

Bulao stood high on a bluff on the left bank of the Bongahon River. On all sides of this village was a magnificent view—of jungle. The Bulao country was swarming with red-shirted *pulajans,* whose rifles commanded that river flowing away into the jungle beneath them.

The Bongahon River was the scene of great activity on the morning of the twenty-first of August, 1904. On its reaches, two Constabulary detachments were operating independently of each other. The first of these was a detachment of Samar Constabulary who had been on patrol. While on this patrol, they had gathered thirty women and children of the *pulajans* and were evacuating them to Tarangnan.

Bowers dispatched these women and children down the river with thirteen Constabulary soldiers, under the command of a Sergeant. In small *barrotas,* which are narrow and easily overturned dugout canoes, the little detachment was gliding along the river. As they turned a curve below Bulao, the party encountered a band of *pulajans* who were swimming the river. The detail opened fire.

On the north bank, a figure rose from the brush and called to the women in the canoes, "Upset the *barrotas* and escape." Before the Constabulary detail could make a move, the women and children had obeyed the command of the *pulajan* chief. The canoes tipped as they rose to their feet and in a moment the Constabulary was struggling in the water, weighted down with rifle and cartridge belt and canteen. The women and children streamed away to the river banks.

The bush became alive with *pulajans*. On the north bank, the force of "Major" Antonio Anugar with twenty-five riflemen and a large force of bolomen, rose from the shelter of the jungle. On the south bank, a mob of bolomen massed to follow the overturned boats that drifted with the current.

It was a hopeless battle. These riflemen on the shore were marksmen, and they were concealed by jungle. Seven of the police detail were shot through the head and the river claimed them. An eighth, shot through the body, sank with his rifle in the deepest part of the river.

The five survivors, attempting to make the best of a bad situation, turned away from the deadly fire and swam for the south coast, where the bolomen massed with shrill cries of "*Tad-Tad*." Three of the five survivors lost their rifles in the scramble to the shore. Second-class Private Delao reached the shore and saw the *pulajan* "Captain "Lucas seize a *barrota* and go in pursuit of fifteen helpless *cargadores* who had been conducting the baggage of the Constabulary. Delao leaped into another *barrota* and went in pursuit of Lucas. He killed the *pulajan* Captain, saving the lives of five of the porters. He was too late to save ten men who were butchered in cold blood by Lucas.

And then Delao returned to the shore, in the face of those armed fanatics. He scrambled up the beach on hands and knees, and in doing so lost his rifle. Private Valentine Buna, the only armed member of the little squad, formed his four mates into a compact spearhead. One rifleman and four privates armed with bolos fought their way through a mass of fanatics and gained the safety of the bush.

Meanwhile, Captain Crockett, with sixteen men of Company A of the Manila Constabulary, was en route to his temporary station at San Pelayo. He was following the besieged soldiers of Bowers' detail down the river. He had stopped at Bulao to investigate the ruins of the burned town and was a half mile above the conflict that was raging below him.

Hearing the sound of rifle fire, he hastened in that direction, and landing at a bend above the scene of the ambush he circled through the tall grass and came upon the outlaws as they were disposing of their loot. Forming his men kneeling in the grass, he fired one volley into the *pulajans* at a range of thirty yards. The outlaws wheeled and Anugar's *pulajans* were quick to estimate the strength of this slender force. They replied with a fire that wounded five of Crockett's men.

Then the cry of "*Tad-Tad*" sounded and Constabulary and *pulajans* charged at the same instant, to bring the conflict to close quarters in the tall grass. The force that Captain Crockett opposed was in excess of 300 men!

Crockett's uniform made him a special target for the *pulajan* fire. He was singled out by "Captain" Francisco Banaldie, who slithered across the grass, slashing and hacking with his huge *talibong*. Crockett emptied his revolver at close range, but Banaldie came on without faltering. For a moment he towered over the crouching Constabulary Captain, bringing down the bolo blade with sweeping, crunching blows. Crockett went down, slashed across the chest and shoulders, and Banaldie, with a last quiver of movement, fell dead across him.

Dead men piled about the prostrate Captain in the center of that fierce dogfight. The bodies of the slain protected him from the blows of the living. The men of the company swarmed above their fallen leader, rallying back to back, to beat off the *pulajans* and extricate their badly wounded Captain.

This fight was one of the bloodiest waged by the Constabulary in the entire existence of the corps. The battle was an epic combat against odds. The sixteen men of Crockett's company gave positive evidence that day of the state of efficiency that had been attained by the Constabulary. The official reports of this engagement give the *pulajan* dead as forty. Actually, more than eighty outlaws were eliminated at Bulao, by sixteen Constabulary soldiers at close quarters.

Among the rifles captured by Crockett were four Krags that had been lost by Company C of the 9th Infantry during that dreadful massacre at Balangiga.

Captain Crockett and Privates Bravo, Fortunato, Figuroa, and Delao were awarded the Medal of Valor on September 14, 1904.

Pedro de la Cruz had suffered a severe loss in this defeat of Anugar's force by Crockett at Bulao, but he was back on the scene immediately. On September 7 he led a force of 300 *pulajans* against Lieutenant Clearman of the 39th Scouts, and was hurled back with a loss of 74 killed.

A year later Lieutenants George A. Helfert and Juan Sulse were able to entrap Antonio Anugar and kill him with most of his men.

As one recalls the campaigns of this period of *pulajan* warfare, one seemingly unexplainable point stands out to require explanation, that being the heavy casualties the *pulajans* suffered in defeat or in victory. Had it been the *pulajans* rather than Constabulary who were the victims of the ambush, it could be readily understood. We would be able to understand these heavy losses more readily if the facts indicated that the *pulajans* had gone against enfilading machine-gun fire or against the fire concentration of a large number of men armed with repeating rifles.

Neither was the case. The *pulajans* were always numerically superior, and they were better armed than the Constabulary. The answer, therefore, is to be found elsewhere: it is to be found in an understanding of three essential facts.

The first of these concerns discipline in action. The *pulajans* were a wild, disorganized mob and their power was largely dispelled before they could close within striking distance of the small, coolly deliberate Constabulary detail.

The second consideration was one of marksmanship. The *pulajans* were inefficient and inaccurate with their superior rifles. Even a Spanish Mauser, or a Krag, loses authority in improper hands. There were among the hillmen a few good shots; some of them were dead shots. The ambush at Bulao proved that. But the great majority knew only the front sight of a rifle. They pointed the weapon in the direction of their quarry, laid the front sight on the target, and pulled trigger. Almost always their fire was high. The Constabulary shot slowly—and made that slow fire count heavily. It required iron nerves in the face of a bolo rush, where every private knew he was faced with the necessity of dropping four or five *pulajans* before they could get within reaching distance with their great blades.

But the most apparent reason for the huge losses of the *pulajans* in battle were the tactics of these religious fanatics. There can be no question of their bravery; and there can be no question of their lack of military judgment. They advanced in massed formation; shoulder to shoulder. They paid small attention to the fundamentals of favorable position, conservation of man-power, or the possibilities of defeat. Worked to a frenzy by unscrupulous leaders who stood aside to watch them die, the *pulajans* believed in the potency of their *anting-antings* against bullets, even as they watched their companions collapse under the accurate fire of the Constabulary.

The *pulajans* sought immediate hand-to-hand conflict with the weapon they understood so thoroughly. That was the horn-handled bolo that was called *talibong,* with crescent shape and edge of razor keenness. When they rushed, it was the responsibility of every Constabulary soldier to support his comrade in the face of that wild attack. The slaughter of Hayt's Scouts and the infantry at Balangiga had proven that fact beyond doubt. When the odds were overwhelming, ten men against two hundred, as was so often the case, the Constabulary were annihilated. Even their accurate rifle fire could not save them then. But even in victory the *pulajans* ringed the dying Constabulary with the bodies of *pulajan* dead.

Apparently the *pulajans,* as did the Moros, sought death in battle. Certainly they did not fear death.

We see them then, a great horde of red-shirted religious fanatics, revolving their bolos in great shining circles; leaping, bounding, hacking; contemptuous of death and possessed with one desire—to come to grips with their enemy. As a man fell, he was replaced by another. Conflict against them was chilling and grim with the threat of death. Required for success were the essentials of discipline and long-sustained expert marksmanship under great pressure. These, the Constabulary developed.

While these fierce struggles for the mastery of Samar had been in progress, the administration was still wavering over a decision in regard to the armament of the Constabulary. The Army was discarding the Krag now in favor of the modern Springfield .30-06. Constabulary officials began an immediate campaign to secure the cast-off rifles for their service.

Although the old Springfields had worn out under the constant usage, nothing was done about re-arming the corps at this time. The records of the various companies show a diversity of arms, ranging from Mauser and Krag rifles to smoothbore Tower muskets and outmoded Remington shotguns. Still unsupplied officially with fighting tools, the Constabulary went about the business of capturing their rifles. A goodly number of Krags had found their way into the corps, all taken from *pulajans* after unequal combat in the jungle.

The quartermaster and commissary departments had improved, but the troops were still wearing the blue shirts of insurgent days, and no satisfactory transport service existed. The Constabulary at this time had acquired a few launches, but they were inadequate to handle the supplies for 7,000 men. In the more isolated stations the men suffered from a scarcity of essential supplies.

The transport division had 780 mules, 172 wagons, and 65 small launches. With men on station at 288 widely separated posts scattered across a thousand miles, this equipment was completely inadequate.

The records of 1904 show the men wearing khaki uniforms supplied at a cost of 64 cents each. There were complaints that the uniforms were inferior (as well they might be, at that price). There was agitation in Manila for a uniform of superior English khaki.

But the morale of the organization was very high, as they went about their work with an almost complete lack of publicity in the United States.

The year 1904 began to draw to a close. Then, on November 10, the *pulajans* achieved a great success near Oras. Here a red-shirted detachment swept down from the mountains and overwhelmed the Scout garrison of the town. Thirteen Scouts were killed and their rifles were carried away. This situation was made serious and truly alarming by other massacres of Scouts which now came to the attention of the Philippine Commission.

If one follows the east coast line of the island of Samar, north to the pueblo of Borongan, he will come to a country as productive as the valley of the Nile River, as mountainous as Colorado, and as wild as Africa. It is the home of the Bisayan deities, "Dia Laon" and "Sed Apa" (in whose names many an innocent life has been sacrificed), and it is one of the bloodiest battlefields in the world.

From the top of the Mesa de Palapag, in northeast Samar, the River Gumay takes its rise, wandering southeast until it tumbles into the Bay of San Ramon. A mile or more from its source it becomes navigable for small boats. A trail from the Catubig River runs along the ridge of the Tatangbang mountain range, the eastern watershed, and a river in some places sixty feet wide parallels it and flows westwardly through a deep gorge. This river has its source on the western slope of the Mesa de Palapag. The Gumay River flows through a valley formed by Mt. Boboyson and the Gumay mountain range, which runs southwestward from the Mesa de Palapag, including Mt. Tabogue. On its southwestern slope on the beach of the China Sea is the *barrio* of San Ramon on the Bay Espiritu Santo. It is reached by a dim path through the thick woods, over slippery ridges, disappearing in the mud of puddles of brackish water in the hollows, and at times into the water courses, then up ladderlike benches onto a flat ridge of sandstone. Here sides of canyons are carved and fretted beyond description, beautiful to the eye but difficult for the combat-laden soldier.

The Cove de Espiritu Santo with San Ramon snuggled at its head is rock-bound, repellent, and difficult of entry, which is possible only during mild weather. White calcareous rocks hard as marble, polished as high as the waves reach at high tide, form precipitous cliffs that project into the sea in a succession of spire-like rocks a hundred feet in height. A peculiar atmosphere of enchantment pervades this locality, whose influence upon the native mariner must be all the more powerful when, fortunately escaping from billows outside and the buffeting of the northeast wind, he suddenly enters this tranquil landlocked place of refuge. No wonder that superstitious imagination has peopled this place with spirits—El Espiritu Santo.

It was the twelfth of December, 1904, and Lieutenant Hendryx, at sea off Espiritu Santo, was proceeding to San Ramon to establish a fighting base. The tragedy of Balangiga was fresh in the minds of his men, as was that earlier massacre at Catubig. The Catubig massacre had preceded the Balangiga atrocity by a year. It had been equally as savage, and twenty-one men out of a detachment of thirty-one men lost their lives, with nearly all the remainder wounded. Three days under the burning sun and through the chilly nights they had fought off the foe, hopelessly, heroically, and without food. Samar owed the government for the lives of these regulars.

With Hendryx aboard the *Masbate* that evening were 198 veteran soldiers selected for their proven worth. The four officers accompanying this command were veterans with numerous victories to their credit, and years of service. Each man was armed with carbine, revolver, bolo, waist-belt, and bandelero, thereby doubling the amount of ammunition carried.

The China Sea, north and south, is very narrow and filled with sand dunes, islands both large and small, swift currents and cross ones, reefs, and deep shallow water, requiring an exact knowledge of his whereabouts by the navigator at all times. Due to the graft of the Spanish Government in making up their maps for use of the Marine, all, or nearly all, were faulty. They had been drawn, it is said, by Spanish Marine officers who sat in Manila and fabricated them without soundings or proper measurements.

The barometer fell steadily, the heat was clammy, the air thick. The *Masbate* furrowed her way in an unmerciful downpour of rain.

Then came a terrific impact, a grinding, crushing tumult. Tons of water fell across the deck and jammed the doors to staterooms. The metallic ring of the laboring engines and the hammering of the waterpumps ceased. There was the sound of a great volume of escaping steam, a smell of hot oil and sulphur fumes, and the engine crew was badly burned by live steam while ascending the hatch.

Portside lifeboats and all but one on the starboard side were disabled. Water on the lower deck ran knee-deep, and waves broke over the upper deck where the command slept. Orders issued that not a man should leave his position were followed, without a single exception, and there was no excitement. The remaining boat on the starboard side was swung in on deck and transferred to the portside—to accomplish this it was necessary to cut away an iron ventilator with axes. The main steam line was severed and the engines disabled. Tons of water rushed from stem to stern. All lights were extinguished. The solid force of the impact lifted the ship out of the water and slammed her down onto a coral reef. The wheelhouse clock registered 5:15 A. M., and they were saddled on a coral reef at the entrance to the Bay of San Ramon, Espiritu Santo, northeast Samar, five miles from the mainland.

With a single lifeline in tow, the First Mate, Tornroth, swimming landward, notified them by pre-arranged signals that he had located a sand dune. A small land-anchor was made fast on the sand dune and the hawser made taut to the bow of the ship. In single file, with guns and belts, but without shoes and leggings, the men slid down the hawser and landed on the dune in water up to their necks. In the course of an hour this very difficult task was accomplished.

With sails rigged on the single lifeboat, the entire force was laboriously transferred, against an outblowing wind, to the mainland five miles across the bay, in command of Lieutenant George S. Holmes. Rapid-fire guns on the *Masbate* were dismantled, and with all ammunition were taken ashore and used in the temporary defense that was hurriedly built.

Lieutenant Hendryx' orders had been to disembark at San Ramon, hike south, cross the Oras River, and proceed cautiously to the River Dolores. There he was expected to encounter an overwhelming force of *pulajans,* armed completely by the two full stacks of arms captured when Company C, 9th United States Infantry, was massacred at Balangiga. He was to make junction with the 37th Company of Scouts who were scouting from the interior along the Dolores River toward the sea, and with the 38th Company of Scouts scouting from the south with the same objective in view.

But Hendryx was not to contact his old 38th Scout Company in this world. As he was disembarking his men in the typhoon that had wrecked the *Masbate,* Lieutenant Hayt was patrolling a jungle fringe not far distant at Dolores. The 38th Scouts encountered 1,000 *pulajans* who attacked on the rear and flanks, and Hayt and all of his company were destroyed. One sergeant only escaped, bearing fearful bolo wounds.

The Scouts went down, ringed with the bodies of the attacking force, for 300 *pulajans* died in that jungle engagement. In that bloody forest the force of Pedro de la Cruz came into possession of forty-seven repeating Krags and a great quantity of ammunition.

Two weeks later, Lieutenant Morton Avery of the 37th Scouts was trapped in the swamps near Oras. *Pulajans* circled the company with steel, and Avery and all of his men save two were killed.

Lieutenant Hendryx found only a silent beach, littered with chopped bodies of men who were to have been his support.

It was disaster that began to awaken the Philippine Commission to the fact that a serious military situation *did* exist in Samar. They were still unwilling to call in the regular army to assist the embattled Scouts and Constabulary. After a conference, they decided to unify the operations of Scouts and Insular Police.

It will be recalled that sometime preceding this period, certain Scout companies had been placed under the command of the Chief of Constabulary. These forces had since been recalled to the Federal service and were no longer available to the Constabulary. Therefore, Samar in 1904 was garrisoned with a mixed force of Scouts and Constabulary, the former being Federal and the latter Insular, and neither able to command the other. It was determined not to renew the process of placing Scouts under Constabulary command. The Commission decided to detail an officer of the regular army to command the joint forces of native infantry. The choice of the Commission for this command was an unfortunate one.

We leave this officer in his undeserved anonymity with the statement that he never once took the field against the *pulajans*, but remained on a coast guard cutter, bawling instructions to his men in the bush through a megaphone. As the "Major with the Megaphone" he remains today to all of the fighting men who eventually conquered Samar.

To continue with conditions in Samar: this officer with the megaphone knew that the Government was very desirous of making light of the *pulajan* activity in Samar. War correspondents were buzzing in Manila, eager to convey to the people of the United States the real nature of affairs in the Philippines. The Megaphone Major was therefore displeased, when he took command, to hear reports of Constabulary officers that *pulajanism* was spreading with the rapidity of a forest fire. We have a record of his remarks when he was informed of conditions in the bush. "Why, there are only a few bolomen hiding in those hills," he said, "Lieutenant . . . . must be getting cold-footed. He should ask for transfer to a cooler district." This about a man whose valor was unquestioned, and who had been in the interior of Samar on intensive campaign.

Under the Major, operations were conducted with a joint Scout-Constabulary force in a manner that was disgusting to the hardened campaigners who had faced many a bolo rush. The puttering ineffectuality of the commander resulted in a series of disasters to American forces, and a great deal of hardship and unnecessary loss of life. The *pulajans* were encouraged to new activity and the good work of months was thrown away.

A word must be said at this time in explanation of the complete inactivity of the Army. We had available in Samar a very efficient force of regular troops who were equipped and ready to fight. Large bodies of beautifully armed and disciplined troops were compelled to stand idly in barracks and watch the struggle the Scouts and Constabulary were waging against *pulajans*.

THE BLOODY ISLAND

The Army acts upon the orders of the administration, and that body was unwilling to admit the presence of an emergency upon the island of Samar in 1904. The commanders of the regular regiments were goaded almost beyond the limits of their endurance at the spectacle of promiscuous raiding within the limits of their authority. *Pulajans* murdered and burned, almost within sight of the army barracks. In Leyte, near Camp Bumpus, the *pulajans* killed men, women and children within a few miles of the 18th Infantry.

The *pulajans* burned fifty-three towns in Samar within a period of less than two months, among them Silonga, which was hardly more than a walk of an hour from the seat of the provincial government at Catbalogan. Lieutenant Bowers of the Constabulary conducted an investigation over this burned region, and estimated that between July, 1904 and September, 1905, fifty thousand persons had been left homeless in the Gandara Valley.

In a report to the Commission in Manila, Governor Feito of Samar sent frantic appeals for protection.

There was no let-up and no relief for the Constabulary soldiers who conducted those dreadful patrols in Samar in 1904. Men went into the bush for a sustained period of service that lengthened sometimes into months of constant battle. They came out gaunt, starved caricatures of men.

But individual heroism was not enough in Samar. Conditions were getting out of hand; there were not sufficient policemen to serve the God of Battles properly. The "Popes" were growing steadily in influence, and new leaders were coming into prominence.

As conditions grew more serious, General Allen came in person to Samar on December 15, 1904, to assume direct command of all Constabulary operations on the island.

Chapter x

THE STRENGTH OF THE HILLS

"Not only roads and trails but also rivers and *esteros* will be patrolled by foot detachments along the banks or by men in canoes . . ."

—*Constabulary Manual*

DRAMATIC, indeed, exhausting the last human thrill, are the stories of small detachments of soldiers who carried the white man's burden on Samar Island.

The trails of Samar were honeycombed with *suyaes* (man traps) and *balatics* (spring traps), equipped with *planquetas* (sharpened hardwood single-pointed spears made of *palma-brava),* with the points smeared with the gum of the *dalit* or *hammaco,* a poison always fatal unless a powerful antidote is given immediately, or the abrasion is cut deeper with a sterilized knife blade. The troops waded through heavy growth of *ticoco* plants which are so active a poison that merely to touch them will produce blisters, itching, and sometimes fever. The mountainsides deeply marked by Nature, serrated and eroded, tumbled into the valley to furnish the path for the mountain torrents which replenished the swift creeks below, their fast-moving current, yellow with mud, thrashing spray high into the air as they tumbled over rocks through the narrow channels.

In these streams the men caught numerous eels. Circling a large boulder in the stream they would stoop down and swing their arms under the side of it and come up with an eel. As it was raised, a native soldier would grab a handful of sand and grasp the eel, at the same time starting for the bank, the soldiers in pursuit taking turns grasping the eel with their hands full of sand, going to the bank where they would kill and skin it.

Dismal, indeed, was the wilderness of Samar. Major Waller of the Marine Corps and a detachment were lost in the jungle for days. Deprived of food and water, harassed by natives, the detachment became separated. Some of the men lost their reason. It is said that one or two of these climbed trees, thinking themselves monkeys, and when rescued refused to come down. As this particular section of Samar was comparatively level land, even climbing trees did not place the observer high enough to see above the labyrinth of brush and forest and give him a sense of direction. Many were the hardships and dangers encountered passing through dark thickets, rugged paths, and narrow defiles in which a small body of men, properly posted, could annihilate the bravest force. The troops waded crocodile-infested swamps or stinking bogs in which the *nipa* grew, with progress almost impossible on account of the mud and the density of the labyrinth of underbrush. With shoes worn out, clothing torn to shreds, and neck, face, and hands sun-blistered and sore; hiking through merciless downpours of tropical rainfall by day, with no relief at night; sleeping on the ground without shelter—such was the lot of the jungle campaigner. With fear always of tarantulas, blood-leeches, scorpions, and centipedes, and with mosquitoes attacking every available portion of the body, and with snakes of the most poisonous kind, Samar was a real "hell on earth."

The *pulajans* had organized themselves into a loose confederacy known as "The *Carzadores* (hunters or mountain men) or *Dios-Dios* (God or any person or thing passionately beloved or adored) of Leyte and Samar," under the command of Colonel Enrique Dagujob, who had fanned the smoldering embers of insurrection into flame. By education, native talent, and cunning, this leader had acquired control of all the people of a large district and had succeeded in eluding the government forces for many years. To fall into his hands as a prisoner meant certain death, perhaps after days of torture of the most savage nature.

In the early 1900's there were not five miles of road or mountain trails on the island of Samar over which horses could be ridden. American troops never had attempted a campaign in a country more difficult or more dangerous than the interior of Samar—it was indeed a rugged wilderness.

By December, 1904, there were 1,800 native soldiers on Samar, and 16 companies of United States Infantry occupying the coast towns. Eleven officers and 197 enlisted men had been killed in action; 48 officers and 991 men had died of disease; 46 officers had been wounded in action; 768 men had been discharged for disability. Firearms to the number of 7,424, and 45,018 rounds of ammunition had been captured or surrendered to the Constabulary; 4,862 outlaws had been killed, and 11,997 prisoners had been taken. From these figures may be measured the quality of the resistance on Samar.

The soul-trying days and nights on almost impassable trails, with men suffering with dysentery, breakbone and malaria fever, Asiatic cholera, and beriberi, are among the most stirring in our martial history. Constabulary officers commanded a large district with half the necessary number of men, and those men poorly clothed, fed, and armed, with no discipline. They assumed the rôle of judge, jury, *presidente,* public health officer, teacher, diplomat, friend and advisor, and jailer; were godfather to myriads of *ninos;* and retained friendly relations with the two religious factions, Roman Catholic and Aglipiano. They were able to throw off disappointment and discouragement; they had an utter disregard of danger of every kind—on the trail, in combat, a cholera epidemic, or a typhoon. They suffered wounds without attention for days. They were half paid, and still they fought.

No medical men accompanied a command on the march or in camp—a vial of aromatic spirits of ammonia, bandages, and absorbent cotton constituted the stock of drugs.

The rank and file of the Constabulary were enlisted in the morning, and, perhaps, detailed to the firing line in the afternoon without instructions in the use of arms, formations, or maneuvers.

Samar is beautiful in spite of its terribleness—volcanic peaks, forests, lakes, open plains, and mountain ranges. It is subject to all tropical ailments. In Samar, even in the shade and with the glorious sea breeze, one suffers with the heat, night or day, whenever he stirs.

Samar has and will have for many years, a primitive, indolent, sullen, and scowling population.

It was into interior Samar that Captain A. E. Hendryx led twenty men on search for Antonio Anugar. As always, before the expedition started, he had the men form a single line, and from head to foot of the line they counted one, two; one, two. In case of an attack all even numbers faced to the right, all odd numbers to the left. In this manner each guarded each other's back. Years of previous service had taught Hendryx to fight in the same manner and with the same arms as the enemy. His company of Scouts had fought with bolos as the insurgents had fought with carbines and the enemy had not relished this at all. Now, as a Constabulary officer, Hendryx applied the same tactics. His advance points were different from those of the American Army. One man was in the trail some ten feet ahead of the next two men, who were one behind the other. The man in the lead had a captured guide tied by a rope so that he could not take French leave, and the guide carried a long *palma-brava* pole with which he jabbed the trail for *suyaes* and *balatics*. From the very start their hike was necessarily slow, for the trail was honeycombed with *suyaes* and *balatics*. Stretched across the trail was coil after coil of *bejuco* (rattan), thickly interwoven and intertwined with thorny vines.

During the rainy season the trails become torpid rivulets of mud—and this was the rainy season. Bordering the trail, and for miles on either side, were magnificent trees of *nara* and *panao,* and *palma-brava* of great hardness. Overlapping the trail was the *palma-areca,* yielding the betel-nut; the beetle, or *buyo;* the St. Ignatius palm *(Ignatia Amara)* that produces the poisonous Catbalogan nut from which strychnine is extracted. Catbalogan, Samar, was named after this tree. All these grew in close proximity to each other.

Arriving at the foot of the hills Hendryx' force was soon making the ascent of the interior mountains under a blistering sun; then over *cogon*-covered hills they went and down between precipitous walls hundreds of feet high; then a succession of hogback ascents and descents through the tropical forest, with trailing plants covered with thorns that tore clothes and wounded the flesh. Finally, the column crossed the watershed between the central cordillera plateau, hiked through dripping *cogon* grass, and emerged from the crisp cool air of that elevated forest-covered plateau onto a projecting cliff, rugged and broken. The sun filtering through the trees cast shadowy patterns of leaf, limb, and trunk at their feet. There were giants of the banyan family with aerial roots forty feet long, securely entrenched in the earth below; and from the depths of the forest came the regular hourly cry of the *kalaos* (the clock of the mountains).

A trail was traced a long way to a projecting, rugged, rock-bound cliff, where it turned sharply to the right. A river snaked its uneven course, replenished by little forking creeks threading their silvery way around hills and through draws; and miles away at the limit of visibility, a placid lake in a clearing, surrounded by bamboo shacks, serenely content—beautiful in Nature's stage setting of tropical shades of green. *Carabaos* and carts proved some semblance of a road existed, and along the river nets were being cast.

At last after an almost perpendicular climb they came out on the well-known mountain, Hurao O Curao. It was a rugged, almost impassable peak, with eroded sides; and there were trails on narrow walls, where a misstep would hurtle one hundreds of feet below to certain death on the rocks. Among the most prominent peaks on Samar is Mt. Curao, of which it is said that no Spaniard saw it save with a pair of field glasses. Every column of Spanish troops that attempted to explore this central mountain section of Samar was out-fought and out-smarted.

At the foot of the mountain Hendryx paused to rest his men. They were treated to another tropical rain that continued unceasingly until dawn of another day. They plodded on then, up the side, and it was here that they made the acquaintance of Antonio Anugar and his cutthroats.

The guide reported that the *pulajans* were advancing from the left side of the trail, straight up an almost perpendicular ascent, and asked if they should fire. Hendryx gave the command for the detachment to fire—it being, at that time, in line of skirmishers. The men held their ground, retained their positions, and continued firing steadily as fast as loading from the belt would permit. Hendryx had fired one shot and had his arm ready to fire another, when he received a wound in his right arm about three inches above the elbow joint. The bullet re-entered on the forearm, passed along between the bones of the forearm and came out on the back of the arm at a point more than halfway toward the wrist.

He collapsed in the high *cogon* grass. When he regained consciousness there were none of his newly recruited men on the scene. He was lying on his right arm and it was with difficulty that he pried his fingers loose from the butt of his revolver. He rose to his knees, and heard a *pulajan* speaking in the Visayan tongue: "There is an American." Upon this, their officer called to the force—some fifteen in number—to advance. The first man to do so came within twelve or fourteen feet of Hendryx before the latter shot him dead. The *pulajan* officer stood some thirty-five or forty feet distant and Hendryx fired at him but failed to hit him. The *pulajans* were endeavoring to surround him at close range so as to use their bolos and daggers. Bent in body, treading high, they came through the grass. In backing up to keep them off, Hendryx fell backwards over a cliff, not recovering until stopped by a fallen log. In falling, he struck his head against the log, rendering himself almost unconscious again.

Knowing that the Pulajans might find him at any moment, he attempted to reload his revolver. He was barely able to do this, his right hand being helpless except for the first finger which he used to extract the empty shells. Although the sounds were confused he could hear the *Pulajans* on the trail above and he judged there were one hundred and fifty (150) of them. Later, upon evidence that they had left the vicinity, he made his way further down the gully to a bench of stone which received the overflow from a ledge above—the constant dripping having formed a bowl. In this he found sticks, leaves, dirt, and a small amount of stagnant water—of this he drank copiously. Most of this day, he reclined on the ground as his strength was almost gone. On the next day, he tried to make some progress and came within sight of the old lookout tower of Erenas. He was able to get within one-half mile of Erenas that night before darkness stopped him and he could go no further. The next morning he tried to get to the town but was so weak that he was only able to travel forty yards in three attempts. He sat down on the ground with his back against a tree and his loaded revolver in his left hand. He could see the trail that he could not travel. He waited patiently for death. He soon fell asleep but later was awakened by the sound of voices. Looking down through the brush toward the trail, he saw Mr. Witte, a former member of "L" Company, 43rd U. S. Infantry, his old Regiment, and three other Americans. He called to them as loudly as he could, "Is that the Constabulary?" Suspecting a trap, Witte and his men faced in his direction as Hendryx called, "This is Hendryx. Come and get me for I am nearly dead."

Today Captain A. E. Hendryx lives in Indianapolis, Indiana, a gallant survivor of the American jungle campaigns. His story, in many ways, is the story of the conquest of Samar Island. He was a fitting member of the American jungle patrol.

Chapter xi

MAGNIFICENT OUTPOST

"To be outnumbered, always; to be outfought, never."

IT was the afternoon of January 6, 1905.

A coast guard cutter, with three officers and 145 men of the Manila Company, Philippine Constabulary, was edging slowly along the jungled fringe of a reef-protected island. It was the northeast coast of Samar, that island of ugly memory that was a graveyard of men.

Captain Cary Crockett was in command of this party; his orders, to establish a post at the town of San Ramon, and from that station to protect the peaceably inclined people of the coast and conduct expeditions into the interior. The boat came to a stop inside a circular bay, and through the green foliage the men could see only the blackened ashes of what had been the town of San Ramon. It had been burned by the *pulajans* and its inhabitants slaughtered or scattered to the hills.

Crockett took three squads and went ashore. There was no sound from the fringe of bush as the force landed and moved away through the high grass. Crockett divided the party into three small patrols; one he sent to the right, another to the left. The center squad he led himself. He was moving through the bush when the silence was broken by rifle fire. Then the distant hills came to life as the *pulajan boudjons,* the great war horns, sounded a summons. There was more rifle fire and a threshing in the bushes and the sound of men in combat. The big red-billed *kalaos*—the horn-bills—took alarm and flapped away with that raucous cry that can be heard for a mile.

The bush opened again after a while, and Crockett came back to the beach. His men were carrying trophies of the chase now; great crescent-shaped blades that were heavily weighted toward the point. The knives were without guards, and the handles were of *carabao* horn and heavily mounted with silver. The edges were as keen as razors. There were the *talibongs* of the hillmen—the great fighting bolos of the fanatical mountaineers. Another soldier carried red tunics, bloodstained now.

They had surprised an outpost of *pulajans* in the high grass.

Crockett had much to consider as he returned to the coast guard cutter at dusk. He had found the town of San Ramon burned to the ground, the civilian population chopped to pieces, and the stores and plantation houses looted. There was no food; nor was there any other mark to indicate that man had inhabited the place, save those few blackened timbers. From papers found on the dead *pulajans* of the force he had routed, he knew that the region was held by one Cinicio Lasara, who was one of the premier fighting chiefs of the famous "Papa" Pablo.

The presence of this strong fighting force in the region had not been anticipated; Crockett's heart leaped as he considered one possible reason for the concentration of this *pulajan* force about desolated San Ramon. It could mean that Pablo's fortress of Maslog was behind San Ramon in the mountains!

For months, the combined forces of Scouts and Constabulary had searched this unexplored island for the stronghold of the "Pope" who was the spirit of the resistance in Samar. Maslog had been freely discussed by the fighting men. Of it they had said, "It's the place everybody looks for and hopes to God he won't find." The officer who found Maslog might go against thousands of *pulajans*.

Meanwhile, Crockett had his orders to consider: "Occupy San Ramon." He had ten days' rations and a fair amount of ammunition. Other than a few cooking pots, blankets for the men, and a medicine kit, that was all he had. Over the horizon was no possibility of relief, reinforcement, or renewal of supplies for an indefinite period, should he elect to remain. And this was Samar, in 1905; a raw and cruel island that swarmed with 7,000 fighting natives.

General Allen had not known, of course, that San Ramon would be found burned and deserted. He had not known that he was sending his Captain into the wild island without hope of replenishing his supplies. Crockett thought of these things. And then he thought of the *pulajan* concentration he had found at San Ramon. This was the place for a detail, he decided; he would stand on his orders and occupy the place where San Ramon had been.

He disembarked his men and that evening the Coast Guard cutter sailed away, carrying Crockett's requisition for more ammunition and supplies and a report of the situation.

There was work to be done. With the severing of this last tie with civilization, Crockett began to prepare for the safety of his men. Three squads were posted on sentry duty at the jungle edge and two additional squads were held in support of the sentries. The remainder of the men began the building of an outpost.

All day they worked, from sunup the next morning until night fell. An area was cleared of grass and brush to permit an open firing zone; a semicircle of rough dwellings was built with a lookout post in a tree in the center. The circle was enclosed with a fence of rough stakes, bound with *bejuco* vines and anchored. It would delay a *pulajan* rush for a few moments.

At the edge of the jungle, tin cans were strung on *bejuco* strings to warn of night attack. The first night in camp the force was not bothered, but they were aware that eyes were watching from the jungle.

The next morning Crockett renewed his preparations for occupation and defense. For several days, the men worked feverishly in the construction of a fort. A bullet-proof, rectangular structure was raised, covered with a steep *nipa* roof. As the roof was highly inflammable, it was lashed in a manner that permitted it to be cut free instantly, should it take fire under attack. Entrance to the fort was made by a single barred gate. A shallow well was drilled and a quantity of brackish water provided.

Then Crockett raised a slender sapling and flew the Stars and Stripes, and at the base of the staff he made a pile of the skulls of the innocent villagers who had died in the *pulajan* raid in San Ramon. His Macabebes muttered as they passed that gruesome reminder of the foemen they were to face.

They were ready now for reconnaisance of the country or for *pulajan* attack. As long as they remained in camp, the jungle was silent, but with every venture of a large foraging party into the jungle, the *boudjons* would boom to indicate that they were under close surveillance.

The days lengthened into weeks in that jungle stockade, and no boat arrived with supplies. The Constabulary were in a desperate position. The food supply failed; they turned to the jungle for supplies. Fish traps were built; the octopi they ate, but not even the hungry soldiers could relish the jellyfish that made up the balance of the catch. One day they caught an eight-foot shark. They feasted then. In the bush were a few coconut trees, and to these expert jungle-men the forest yielded bamboo shoots, palm cabbage, *gabi* roots, from which *poi* is made in Hawaii, and a huge root called *palawan* which Hazzard said tasted like "army issue soap."

They ran out of tobacco, using papaya leaves, ground and dried, as a miserable substitute. Salt they evaporated from the sea water, and they found wild honey in the hollow trees of the forest. As the food supply grew scantier, Crockett and his Lieutenants, Hazzard and Mann, adopted the dangerous expedient of stealing from the camp at night to sit in trees in the dark, in an attempt to shoot wild boar. Occasionally they were successful, wondering each time if the flash of their rifles would bring a horde of *pulajans* upon them in the dark.

With 147 men to feed, the search for food became feverish. One day Crockett crossed through the swamps to the higher ground in the foothills and came upon a troop of monkeys. He shot as many as he could, and they were a welcome change of diet.

Again, they ate *kalao,* the huge horn-bills that are remarkably good eating—when they can be killed. During these starvation days, Crockett had occasion to marvel at the fortitude and stoicism of his little Macabebe soldiers. He writes, "I never ceased to wonder at the constant cheerfulness under conditions of service severe enough to sap the spirits of any body of men. The men were on continual campaign, with death in many painful forms ever lurking in the background. Discipline was strict, if not harsh, the pay was small, the clothing and equipment inferior, and the food poor even under ordinary circumstances; and yet, they not only re-enlisted, but there was a waiting list of friends and relatives to fill vacancies caused by death and disease. They were ever ready to follow, or to precede white officers into any danger, blindly and without question, so long as they realized that their officer was there to guide them and direct them. In their devotion and eagerness to please, they resembled a well-trained pack of hunting dogs."

We see his Macabebes there, at lonely Fort Defiance, with their orchestra of many pieces playing banjos and native wind instruments in the very face of starvation and death.

One day Hazzard went fishing. He returned with a boatload of immense fruit bats, repulsive creatures with thick, brown fur. "What did you use for bait?" Crockett said. Hazzard grinned, and the natives went about with the preparation of the ghastly creatures. They were strong of flavor and rank of odor, but officers and men ate them, with enthusiasm, if not with relish.

Hazzard was a forager of renown. Another day he came back with two twenty-four-foot pythons. "Forty-eight feet of snake," he said, "and every foot good meat except the heads." The company fed royally upon one of the snakes. The other they cut into strips and dried in the sun for emergency rations.

Then they found a reef with fine *taraquito* and the big saw-toothed barracuda. These they caught and cured. Every day a boatload of men patiently fished the reef with varying success.

The failure of the *pulajans* to attack became a source of worry to Crockett. As the weeks lengthened, he decided to take a force into the bush to seek the *pulajans* who would not come to him. Eighty of the men in the best condition were detailed to accompany him; the remainder were left to garrison Fort Defiance under Hazzard. Lieutenant Hazzard received his orders: he was to hold Fort Defiance against all attack and never for a moment relax his vigilance. If Crockett failed to return from the jungle, he was to continue to hold the place until relief arrived from Manila.

At two o'clock in the morning of the scheduled departure, Sergeants Bustos and Alalay were sent out to reconnoiter the trails and to attempt the capture of a guide. The main body left the fort at half-past four and met their scouts a short distance from the post. The Sergeants had no guides, each claiming that the man they had stalked had refused to surrender and was therefore killed. They had crept upon a *pulajan* outpost and used their long knives silently.

Crockett pushed on through the bush. He sent two squads ahead to develop any enemy patrols. The advance was through great canyons of granite, across mountain streams, over giant boulders, and around sheer mountain spires. The crest of a watershed was scaled in a chill rain that obscured the view. They crossed swamps and waded formidable mountain rivers which threw spray high on the sides of the rocky cliffs. After a march of fourteen hours, they made camp on the crest of a rounded knoll. There was no attack.

The next morning they resumed the march, over another mountain range and through the thick black muck of the lowlands. They came at last to a burned hut and a fork in a native trail. They turned along the most used trail and began to pass cleared areas and patches of *camote* beds. Towards evening, the company descended into a wide, open valley, where *cogon* grass grew ten feet in height on either side of the trail. Here the trails were mere tunnels through the grass. Crockett went ahead, crawling on hands and knees through the high grass. He emerged upon a hard, packed trail that was almost a road.

He took this trail, and at a bend came face to face with three *pulajans*. Each was in uniform and had two bolos and a dagger strapped to his waist. Crockett spoke to them, ordering them to lay down their arms. The man in front made a motion to comply; then, with a swift movement, he slashed at the American officer. The second *pulajan* tried to stab Crockett in the side as the third rushed with a bolo.

Sergeant Alalay saved Crockett's life. He killed first the leader and then the second man, with charges of buckshot. Crockett dropped the third, in mid-air, with his revolver.

The presence of this broad trail indicated to Crockett that a considerable *pulajan* settlement was somewhere ahead in the mountains. It also indicated the inadvisability of dividing his force at this time. No further *pulajan* activity was developed, and in the middle of the night Crockett turned back in the direction of Fort Defiance. He had the information he needed for concerted attack. The next night, after a terrible forced march across the mountains, he was back in the stockade. He sought his blankets and had hardly more than fallen asleep when he was awakened by the heavy roar of rifle fire.

The *pulajan* attack on Fort Defiance came in two great waves. In the first advance were sixty natives, carrying long poles lashed with burning torches. They were supported by heavy rifle fire from the jungle edge. Their object was to fire the grass roof of the fort and to force the occupants into the open to be chopped down by bolomen. The leading wave dashed in and leaned their torches against the walls. The roof went up like tinder and the scene became as light as day. Then the fanatics swarmed against the stockade walls, each with two bolos lashed to wrists. Up they mounted, lined in the bright glare of the burning roof.

As Crockett cut the lashings supporting the burning roof, the heaviest wave of attackers struck the stockade walls. The roof collapsed upon them and for a moment the rush was halted as men burned to death at the foot of the walls.

Then the main wave of the *pulajans* came, bounding across the open space, shouting the battle chant of "*Tad-Tad.*" Their red uniforms glinted in the light and the long white capes fluttered in the breeze as they came. Barearmed and barelegged, they rushed to the attack, each man swinging his bolos in circles about his head.

All the time the attackers were covered by a tremendous, if ineffective, fire from the bush. The *pulajans* had hundreds of rifles in this engagement. On the stockade walls, the Macabebes gathered in compact, supporting groups as the fighting became hand to hand. The *pulajans* who survived the Constabulary fire scaled the fifteen-foot walls as easily as monkeys. A wave of dead men piled up against the foot of the walls. At close quarters, the revolver and rifle fire of the Constabulary was deadly. But the *pulajans* came by the hundred, careless of death, eager to gain those walls to the interior of the fort. The Macabebes were demons on defense. In the press of battle, two of them watched a *pulajan* scaling the wall in front of their position. "You hold him," one suggested, "and I'll stab him." Which they did.

The smoke of black powder eddied about the fort as the fight roared through the night. The dawn came suddenly, to show the *pulajans* gathered at the jungle edge, chanting their prayers. They made a last, frantic rush, which was stopped by a volley. Whatever their distorted convictions, the *pulajans* were brave men.

This action was one of the greatest disasters to the *pulajan* cause in the history of the fighting in Samar. The *pulajans* learned that day that they could not attack Constabulary in barracks. When the police emerged from Fort Defiance, they found the bodies of a hundred *pulajans* on the field. There were bloody trails in the forest where others had crawled away to die. The less seriously wounded had been carried away. Seven hundred *pulajans* made the attack, and the band of Anugar, the terrible, was disorganized and broken.

The dead were buried in the soft sand of the square, two deep in a long trench. Practically all of the Constabulary were wounded but the casualties in killed were few. The *pulajans* had not been able to get within reaching distance with their *talibongs*. Their attack had been a useless, futile gesture. The Constabulary had killed them to live.

This action was on February 23, 1905.

Among the prisoners captured in the attack on Fort Defiance had been a small boy named Feliciano, who had marched bravely to the assault with the men. He attempted to stab Crockett when picked up on the field. However, kind treatment and gifts changed his allegiance, and he became an adoring worshiper of the Constabulary Captain.

From this small *pulajan* Crockett was able to gauge very accurately the extent of the religious movement. Feliciano had been with Anugar's band, and he was able to recount the attacks on Catubig and to tell of the massacre of Lieutenant Hayt and his company at Oras. It became apparent to Crockett, as he listened to the small boy, that his troop was one of the few left on Samar. He was still faced with the prospect of virtual starvation, and his supplies of ammunition were scanty. He had received no word from Manila, either in orders or supplies.

Making the best of a very bad situation, Crockett decided to leave all of the men but sixty at Fort Defiance, under the command of a capable Sergeant. He would take the sixty men, with Lieutenant Hazzard, in an effort to break through to the south to some occupied village, and there secure supplies for his men. He intended to return to the fort with supplies via the sea. He was convinced that the severe defeat the *pulajans* had suffered would discourage further attack on the fort during his absence.

The officers and men who began that march to the south were little more than skeletons. Hazzard became delirious and had to be carried. The strongest of the men walked supporting the weak. Crockett kept his feet, moving from one end of the column to the other, driving his men through a wilderness that threatened to engulf them.

Through that somber forest the little force staggered day after day. They ate roots. They killed the snakes in their path and ate them. They were truly men against the jungle. Every night, they built their small fortified camp, even as they drooped with hunger and fatigue. They plumbed depths of weariness that are revealed to few men. They reached deeper and deeper for that hidden store of endurance that was seemingly inexhaustible. They kept to their feet —long after they should have stretched in the swamp mud to die. They proved to themselves how really tough and fibrous the human organism can be.

After days of bitter marching, they came to the edge of a broad stream that flowed away through the jungle to the sea. Their uniforms were ripped and tattered; their flesh was torn by thorns and their hair was long and matted. They were almost naked and they were starving. They had reached almost that last limit of their endurance.

As the gaunt, battered men rested there by the river bank they heard an incredible sound—the sputter of the engines of a boat and the swirl of propellers. Even as they rose to their feet an army launch came around the bend, filled with blue-shirted regulars. Crockett hastened to the bank and waved his hat as the men cheered. The soldiers on the launch snapped to action. The launch sheered away and a stream of bullets from a Gatling gun spouted the mud where Crockett stood.

Crockett's battered force had been mistaken for *pulajans!*

In the face of this great disappointment, there was one ray of hope. The presence of the army launch meant that down the river was a garrison of regular troops.

The men set about the building of a raft. After hours of work, they felled a great hardwood tree with their bolos. Laboriously, they hauled it to the river edge. In the water at last, it sank like a stone!

They turned then to the porous stalks of Manila hemp plants. They were nothing more than a vegetable pulp, but this was within the limits of their strength. A float 100 feet in length was constructed and launched. The river was tidal, and they were forced to crouch in the rain for six hours, until the current flowed seaward again. At midnight this change occurred, and they boarded their strange craft and set off down the stream.

The river was alive with enormous crocodiles which followed the raft hungrily. At six-hour intervals, the party sought the land and waited for the changes in tide. There the sick were kept alive with a handful of rice soaked in water and eaten raw. All attempts to build fires with powder from a shell or with flint were unproductive of result. The woods were sodden with rain.

After two days' journey in this manner, the river banks changed in character, to develop into high rocky canyons. The sides of the canyon, where the current ran swiftest, had eroded into shadowy caverns, and the water-logged craft was drawn into these as it was swept by the current. The raft began to break up as it collided with sharp outcroppings of rock. The presence of giant crocodiles added to the danger.

Many years later, Crockett stated that of all his Philippine experiences, the night in these caverns was most harrowing. The splash of the crocodiles about them; the rank, musty odor of the brutes, and the added possibility of death by drowning combined to make a night of terror.

At daybreak of the third day, the pulpy mass of vegetation began to disintegrate and Crockett landed his men on the shore. Lieutenant Hazzard and ten of the men had reached their final limit. They were unable to walk.

Crockett selected twenty of the strongest of the men and set out to find the post. Before he departed, he cautioned the men to remain together and be prepared to resist attack. He found a trail, overgrown with vegetation and swarming with leeches to make marching a horror. Every valley was a slough of mud, through which the men pulled themselves with difficulty. After ten hours, they came to a rice field, and in the distance they could see the walls of the ruined town and the sea.

It was Oras, scene of the Scout massacres of Lieutenant Hayt and his company. The town at one time had had a population of 10,000, but on that day not a building had escaped the *pulajan* vandalism. It was a mass of burned timbers. But the stone walls of the church were standing, and above those walls Crockett saw a sight to gladden his heart. It was the flag of the United States.

The accumulated miseries of two months were forgotten as Crockett led his ragged force into the town. The soldiers on sentry duty saw him and ran to the *cuartel* with a cry: "*Pulajans.*" The bugles blew "Call to Arms"; riflemen hastened to combat stations and the snout of a Gatling gun pushed through a crevice in the walls.

Crockett halted his men at the edge of a barbed-wire entanglement and made himself known. He expected a bullet at any moment. An officer and twenty men came from the church to inspect the Constabulary detail, and in a few moments Crockett was in the presence of the commander of the regulars.

There he was refused the loan of the government launch to pick up his men on the river bank in the interior. A junior officer of the regulars who had volunteered to go for the party was refused permission to do so by this crotchety commander of the regulars. Today, the actions of this officer surpasses our understanding; his own officers were disgusted with him. He was not a representative specimen of the regular officer in the Philippines.

When the tide changed Crockett secured native canoes, and at noon of the next day Lieutenant Hazzard and the remainder of the Constabulary were safe at Oras.

As Crockett came back with his men, a government steamer came into Oras and General Allen disembarked to take command of Constabulary operations in that vicinity.

The meeting between General Allen and Captain Crockett was perfectly timed and intensely dramatic. Crockett was miles from the station to which he had been assigned. Allen was certain to ask questions.

Constabulary captains do not talk of their hardships or question decisions when reporting to the commander after an absence of two months in the bush. Nor did Crockett; he launched instead into a report of the building of San Ramon fort and of his first trek into the hills in search of Maslog.

General Allen, the courtly, became Allen the military personage on the instant. If Maslog had been discovered, why had not Crockett attacked after defeating Anugar—that was what Allen wanted to know.

Another sudden thought crossed Allen's mind as he gazed

at his young Captain. "Aren't you a long distance from your assigned station, Captain?" he said.

Crockett was forced, then, to tell Allen the story of that miserable two months of starvation at San Ramon, where he had apparently been forgotten by a commanding officer. The General unbent as he listened to that story of jungle outpost duty, and then, for the first time, Crockett learned the reason for his seeming abandonment at San Ramon.

It had been the "Megaphone Major," previously mentioned, who had been responsible for those terrible privations at San Ramon. He had ignored direct orders from Manila to come around to the east coast of Samar with supplies and take the field in co-operation with Crockett's company. As an excuse, he had stated that the monsoons were so bad that boats could not navigate.

After Crockett had completed his report to General Allen, the immediate organization of a concerted attack on Maslog was undertaken. Captain Todd, with a column of the troops from Oras, proceeded up the river to the head of navigation and established a post at the burned village of Concepcion. Ballard's company of Scouts took station there with the mission of clearing the country of *pulajans* and protecting the returning villagers from future raids.

The Scout companies who had arrived from Manila with General Allen, commanded by Captains Nickersen and Cook, were to collaborate with Crockett's detachment in an expedition against Maslog.

This force was embarked on the Coast Guard cutter and removed to San Ramon, where Allen made plans to lead the assault in person. It was a great relief to Crockett when the ship rounded the point and he saw his ragged soldiers lining the parapet to welcome his return.

The old Macabebe Sergeant was a soldier to the end. When he saw the Brigadier-General's flag on the cutter he formed his company under arms. When the General landed at the little post the garrison paraded, rifles were snapped to the "present," and there was the prescribed flourish of the homemade musical instruments as the General drew near.

And then Allen stepped out in front of that hungry lot and thanked them for the soldierly qualities that they had displayed and told them that he was proud to be their commander.

The first thing that Allen did, when the men had been fed, was to call the little *pulajan* boy Feliciano for interrogation. The little boy told the General that he could lead the force to Maslog and he gave the first complete description of the *pulajan* rites.

He told of the torture and the killing of prisoners by boys of his own age. He told of having officiated at a *verdugo* himself upon several occasions. The *pulajan* prisoners were tied to stakes and the little boys were shown where to strike. Once it had been hard, Feliciano said, when the victim had been a woman, whose cries and struggles had made him think of his mother, who had been killed also by the *pulajans* when Feliciano had been adopted into the tribe.

The *pulajans* included as a ritual this compelling of children to act as executioners; it was believed to make them bloodthirsty and fearless in battle.

The little boy had never seen "Papa" Pablo; not even Anugar, the war chief, was allowed admittance to the presence of the "Pope." He was an awful figure who was shrouded in mystery and was the possessor of supernatural powers.

After listening to Feliciano, Allen decided to move with 200 Scouts and Crockett's company, with the exception of seventy men who were to be left at San Ramon under Lieutenant Hazzard. The original destination was to be Cagamotan, on the north coast; from there, they would march on Maslog.

Every Macabebe in Crockett's company clamored for the opportunity to join the attack on Maslog. The men who had remained at the fort maintained it to be their turn, while those who had accompanied Crockett insisted that the assault should be their payment for the miserable trek through the jungle. Crockett picked the men by lot, eliminating only the obviously unfit.

At daybreak the next morning the troops disembarked from the cutter at Cagamotan. As they landed, a *boudjon* boomed in the distance and the sound was taken up and repeated until it had died away in the far hills. The column was formed and began to move up a well-beaten trail that led away into the interior.

Crockett led the advance into the bush; behind him, eight Constabulary armed with repeating shotguns. Then came General Allen, followed by the balance of the Constabulary. The two Scout companies formed the rear guard.

Allen was in the full dress uniform of a brigadier-general; Crockett was gaudy in Captain's bars and glaring red epaulets. At Allen's insistence, they were going into action attired as officers and gentlemen. The General had refused to take a less exposed position in the center of the column; this splendid Kentuckian was too dignified to be afraid.

Orders were passed along the line that when the ambuscade came, the men were to rally in platoons and fire at will. They were to ignore the rifle fire of the *pulajans* and concentrate the fire upon the bolomen.

The trail traversed a gentle slope, covered with patches of *cogon* and jungle. They moved ahead slowly. The blare of the *boudjons* had ceased; it was steaming quiet in the jungle. No enemy was seen or heard, and the mental strain became greater as they climbed in the foothills.

For two hours they mounted steadily into the hills. Then the head of the column came to a level place in the path where the grass grew high and thick on either side of the trail.

A shrill whistle blast sounded almost in Crockett's ear and he spun in his tracks to fire at a red-garbed figure that rose beside him. From either side of the trail a blast of rifle fire was loosed at the Constabulary. As the flame from the rifles spouted in the very faces of the advance detail, Crockett felt a sledgehammer blow and the shotgun dropped from his hands. His left forearm had been shattered by a heavy caliber soft-nosed bullet.

The cry of "*Tad-Tad*" went up, and swarms of red-coated bolomen rose from the grass on either side of the trail to rush the Constabulary. Crockett was attacked by four natives, each eager to kill the white officer and gain possession of those red epaulets. Half of the men in the two leading squads were down; the remainder were hand to hand with the *pulajans*.

Twenty feet in the rear General Allen was maintaining his reputation as a wing-shot. He dropped three *pulajans* with as many shots; in a moment, he too was hand to hand with the mountaineers. Crockett had stopped three of his charging adversaries with his revolver. As it clicked on an empty chamber, Sergeant Alalay, the faithful, turned from his own troubles and dropped the fourth with his shotgun. Crockett was unable to reload his weapons; with his left arm spurting blood, he seized a bolo and defended himself against the rushes of the *pulajans*.

The men were spread out in single file along the trail and the entire column was under simultaneous attack. First Sergeant Bustos, a very gallant and brainy fighter, saved the situation and the lives of General Allen and Captain Crockett. He closed his men up and faced them alternately, right and left, delivering rapid-fire volleys until the waves of *pulajans* fell back with the ground littered with their dead. Bustos then rallied his platoon in a half circle and advanced through the grass, killing the *pulajan* riflemen almost to a man.

Crockett fainted from loss of blood as the attack ended. He was revived and prepared for return to the coast guard cutter. He objected so strongly that the General permitted him to remain with the party.

The advance moved on, Crockett still in the lead. His wrist and hand were without feeling. Sergeant Alalay remained at his side, steering him around the pitfalls in the trail and the spear traps that grew more frequent as the stronghold was approached.

They came out on the top of the mountains at Maslog, the mysterious. It was to be the lot of no white man to see this *pulajan* fortress that had been the subject of so many uneasy rumors. The defeated *pulajans* had retired swiftly and burned their fortress to the ground.

With the picture of Cary Crockett still leading that column as they retraced their way along the trail to the seacoast, we usher this valiant character from the pages of this chronicle.

He fought *pulajans* for one more year, following his discharge from the hospital. In his jungle service, he acquired three grievous wounds. In 1906 and 1908 he was in Cuba, with the American provisional government; and on January 23, 1908, he resigned as Captain of the Philippine Constabulary to accept a commission as Second Lieutenant in the regular army. Today, he is still on active service, with the rank of Colonel.

Chapter xii

THE REGULARS IN SAMAR

"When a small detachment is encamped in a country infested by large ladrone bands so that a night attack may be feared, it is advisable to go into camp before nightfall, cook and eat food, build small huts and make arrangements for remaining during the night. A short time after dark the patrol should then slip quietly away, leaving the camp deserted with small fires burning . . ."

—*Constabulary Manual*

WITH Samar in flames, the course of the Philippine Commission was well pointed out. No longer could the distasteful idea of intervention on the part of the official armed forces of the United States be delayed. Men of the Constabulary and Scouts were dying as the Army sat watching the struggle. It was time that the best armed troops in the Islands took over some of the responsibility of putting down the revolt in the flaming island.

But even in this bloody crisis, while men were dying, the interests of politics had to be served. Nothing was done to relieve the situation in Samar until four days after the Presidential election in the United States. It was the fate of the fighting men to await the pleasure of the politicians. On November 12, 1904, Governor Wright went to Samar, and upon his return the 700 Scouts and Constabulary on station there were increased to 2,000, and 1,600 regular soldiers of the 12th and 14th Infantry were thrown into the area.

On December 29, General Corbin, at the request of the Governor-General of the Philippines, ordered out regulars to garrison the towns of Llorente, Oras, Taft, Tarangunan, and Bulao. With this force in garrison, the Constabulary took the field in the interior in an effort to stem the *pulajan* growth.

Allen remained in Samar to lead his men, but after five months of vicious campaign he found the situation too severe for the limited troops of the Constabulary. The Insular Police could not be reinforced without drawing too many men from other badly pressed stations elsewhere in the archipelago.

A new leader had arisen among the *pulajans*. Dagujob was his name. He had first come to prominence in Leyte, where he had been known as Enrique Villareal. When conditions became favorable in Samar, he had migrated thither and established strong positions at Maslog on the east coast and at San Jose on the west. In later collaboration with "Pope" Pablo he became the most formidable leader on Samar.

On January 8, Dagujob met Lieutenant Averill of the 37th Scout Company and Lieutenant Helfert with fifty Constabulary in a short fierce battle at Maslog, that resulted in the retreat of the *pulajan* to San Jose. At San Jose, Dagujob consolidated his force with those of Pedro de la Cruz and "Papa" Pablo. Seventy-eight houses were built within a newly constructed fort.

Here Dagujob was struck again on February 3 by Lieutenants Cook and Overly of the Scouts, with 100 men of the 18th and 35th Companies. Hastily Dagujob recrossed to Maslog, where he was again attacked by Lieutenants Wilson and DeCourt on the eighteenth.

The reports of the encounters show the difficulties of these jungle campaigns which were so often indecisive: "The attacking forces in each of the three cases, although successful, were unable to push their victories through. The long marches required to reach these places, the ignorance of the nature of the defense, the denseness of the surrounding jungle, the comparatively small columns, and the responsibilities caused by men early killed and wounded in the actions brought about the withdrawals."

During the year 1905, a murderous *pulajan* named Teducduc made his appearance in the Gandara Valley, and Constabulary detachments were in the field constantly in an effort to protect the civil population who were subject to the raids of this leader. Teducduc was elusive and difficult to find; for many months his principal retreat in the mountains was not found.

Finally, on May 17, a detachment of fifty-five Constabulary from the post near Mugtaon made an expedition to the north, and after passing through a region of cultivated fields that had been previously unexplored, they located the fortress of Teducduc and burned it after a savage fight. This action stirred up the lawless elements, while the discovery of the large fields of food crops was a severe blow to the *pulajans,* who were dependent upon the fields for their food supply.

In revenge for the assistance given the Constabulary in this campaign by the citizens of the *barrio* of Santo Nini, the *pulajans* assaulted and burned the town on June 2, slaughtering thirteen men, women, and children. Vigorously pursued by the Constabulary, they reappeared on the opposite side of the island at Balangiga and Basey, where they burned and murdered under the direction of Dagujob, Teducduc, and Pedro de la Cruz.

Farther south, Anugar and Aguilar were active; the pursuit of these five leaders was more than the limited strength of the Constabulary would allow. In spite of constant patrol, the *pulajans* were getting out of hand rapidly. On May 21, 1905, General Allen was forced to telegraph the Governor-General, recommending the division of Samar into two sections, the east district to be turned to the army, with the Constabulary retaining the west coast.

This was done, and the Constabulary withdrew all garrisons from the army territory and concentrated their forces for a final drive on the west coast.

No martial law was proclaimed in Samar, but within six months seventeen Scout companies had been ordered to the field, as well as four troops of the 8th Infantry under Colonel Smith and the entire 21st Regiment of United States Infantry.

The Scout companies and Constabulary detachments, withdrawn from other islands and sent to Samar, gradually raised the combined force of native troops to about 1,700 or 1,800 men, but the results of the first half-year's campaigning were not wholly reassuring. The Constabulary had gradually become involved in field operations with native troops equal to a small brigade in numbers; this in one of the largest and most sparsely populated islands of the archipelago, and one which is without a single road or practicable trail for pack animals. For such a campaign the Constabulary organization was unprepared, by reason of lack of staff and supply departments, a deficiency which could not be remedied by expedients. The insufficient means of communication with the east coast constituted a very embarrassing factor throughout the period of operations.

Scouting with an insufficient force for an elusive and half savage enemy, through virgin forests and dense jungles; wading in water courses, tortured with leeches, and dependent upon native carriers of doubtful loyalty, it is not to be wondered at that the campaign was long and unproductive of brilliant actions. To add to the difficulties, the enlistments of both Scouts and Constabulary began to expire at the height of the campaign; this, with the losses in action, combined to disorganize operations and fill the ranks with recruits, many of whom had never fired a gun. The enemy had cause for elation in the success with which they had frequently met detachments of native troops, and the facility with which they had escaped whenever the tide turned against them during more than five months of campaigning.

Whatever may have been the original cause of the outbreak, it was soon lost sight of when success had drawn a large proportion of the people away from their homes and fields. The lawless bands degenerated into opposition to all control, and carried on a reign of terror throughout a large portion of the island. It became simply a question of joining the *pulajans* or being harried by them. In the absence of proper protection, thousands joined in the movements to the extent of rendering aid both by furnishing food and giving information of all movements of troops. This led to the ambushing of detachments, and forced the sending of much larger columns of native troops than had previously been necessary. The movement found no permanent lodgement in the southern part of the island, nor had the northwestern part been seriously involved.

The situation gradually got beyond the control of the Constabulary and Scout forces, and in order to free them from garrison work in the towns, sixteen companies of the 12th and 14th Infantry were distributed about the disaffected coasts to enable the people who so desired to come from their hiding places and undertake the rebuilding of their burned homes.

Under the protection of the troops much progress was

made, but there still remained the unsatisfactory fact that the people feared to engage in any profitable occupation out of sight of the villages, for fear of the *pulajans*.

Under later authority of the Division Commander, the regular troops were directed to co-operate with the Constabulary forces whenever requested by the Chief of Constabulary. Several expeditions were made, but contact with the main body of the enemy was not effected, owing to lack of information and to the fact that all of the natives of the interior appeared to be friendly to the *pulajans* and acted as outposts for them.

The effort to use the Army to assist the Constabulary to preserve order, without formally authorizing the troops to assume active and independent operations under military control, was not productive of the most efficient and economical results. The drain upon the Insular Treasury was growing to considerable proportions, and the end of the campaign was not in sight after nine months of constant and severe field service.

During the month of May the necessary authority for troops to aid in quelling the disorder, acting under the direction of the Department Commander, was received. Immediate steps were taken to put in the field a number of strong detachments, each comprised of half a company of American troops and a half a company of native troops; and arrangements were made so that upon the return of a detachment to its base station, another detachment should be ready to start immediately.

During the progress of the campaign with native troops the *pulajans* had developed a systematic method of attack, which consisted in placing bolomen in ambush in high grass *(cogon)* on both sides of a trail, and when a column got between them, another squad of *pulajans* armed with rifles would open fire from the front. During the confusion of deploying to the front, the two lines would rush from the flanks. This mode of attack proved so successful that it became necessary to provide for it. A special drill was devised which did away with advance guard and flankers, the detachments marching in columns of files well closed, and numbered alternately one and two from front to rear. At the first sign of an enemy, all the numbers one faced to the right and the numbers two to the left and opened fire into the moving grass whether the enemy could be seen or not. When the flank attack was repelled, rapid fire to the front was taken up by those at the head of the column. The first attack of this kind on American troops, a company of the 21st Infantry, resulted in a loss of nine of the *pulajans* killed and twelve wounded, four of whom subsequently died, and without any loss to the troops.

Within ten days after the army was called upon to act, under the orders of the Department Commander, important captures were effected and information secured as to the location of the camp of the real leader of the *pulajan* movement, Colonel Enrique Dagujob. By education, native talent, and cunning, this leader had acquired control of all the people of a large district, and had succeeded in eluding the government forces for nearly a year, except when he planned attacks or ambush. He had assembled as his immediate following about fifty riflemen and an average of two to four hundred bolomen. Captured papers written by him constantly bore the heading which gave the title of the *pulajan* forces as "The Army of Cazadores (hunters or mountain men) of Leyte and Samar."

As soon as the information came into the hands of Major H. A. Leonhaeuser, 21st Infantry, commanding the troops in the Catubig Valley, a detachment of eighty men of Company E, 21st Infantry, and the 38th Company Philippine Scouts was organized under the command of Captain Cromwell Stacey, 21st Infantry, Captain W. W. Taylor, Jr., Philippine Scouts, commanding the Scouts.

Captain Cromwell Stacey, now a Colonel, retired, of the United States Army, living at Port Angeles, Washington, was, without question, one of the finest soldiers developed by the Philippine jungle wars. No officer of regulars, Constabulary, or Scouts saw more service against the enemy than this steel-nerved officer who saw intensive service against *insurrecto, pulajan,* and Moro. No officer in the jungle patrols had greater stock of bush lore than Stacey, and few were his equal in the screened movement of troops in the jungle. He was a fighting man who made each campaign a *planned* campaign, and he overlooked nothing to insure the success of the operations or the safety of his men.

With Crockett, Stacey stands in the very forefront of that gaudy company of brilliant fighting men that were the product of the jungle wars. Years later, the two served together in France, where Stacey topped off a brilliant and lengthy fighting service that embraced four wars as the commander of the so-called Lost Battalion of the Argonne Forest. In the Argonne Forest, as in the Philippine jungle, Stacey fought a planned battle. The men of the miscalled Lost Battalion were not lost; Stacey knew where they were at all times, and it is interesting to note that the war records of 1917 show that the men took up that exposed position in the Argonne against the wishes and the judgment of their commander. A General Order sent those men into an impossible position—not Cromwell Stacey.

On May 31, 1905, when Cromwell Stacey took the field against Dagujob, he was a young Captain of the 21st Infantry and a veteran of the northern island campaigns. It was fitting that he should be selected to conduct operations against Dagujob for, during the earlier days of the Philippine Insurrection, Stacey, as a First Lieutenant of the 19th Infantry, had fought Dagujob who was then a Captain of artillery of the insurgent army.

On the afternoon of June 1, 1905, at Camp Hartshorne in Laguan, Samar, his final instructions from Major H. A. Leonhaeuser of the 21st Infantry were received, and he embarked with his men on the launch *Florida* for Catubig, Samar. He had aboard thirty-five men of Company E of the 21st Infantry, and one Hospital Corps private. Upon arrival at Catubig, Stacey organized a disciplined force of 120 *cargadores* (bearers) into six gangs of twenty each, each marked with strips of colored calico tied around the wrist. There he was joined by forty-four Philippine Scouts of the 44th Company under Captain W. W. Taylor, and with two native guides Stacey set out into the bush.

The party was accompanied by Third Lieutenant Juan Sulse of the Philippine Constabulary, who collaborated in the securing of the guides and was permitted to join the attacking force as an observer. By eight o'clock of the evening of June 2, the party had crossed the Catubig River and had taken up the trail formation that was preserved during the entire campaign. The elements were arranged as follows: First the guides and a few Volunteers, immediately followed by Captain Stacey and a part of Company E of the 21st Infantry; then Captain Taylor with a portion of his Scout force; then the *cargadores,* who were followed by the rear guard of Scouts and Infantry under First Sergeant Joseph Lees. The rear guard numbered about twenty men. The column marched entire, that is, there were no divisions into advance, main body, and rear guards, as is usual in civilized fighting. Stacey had learned from many a bolo rush that in the Philippines it was better to keep the men closed up and well in hand.

The column remained on the main trail until within five miles of the *barrio* of San Vicente, at which point they turned away to the right, crossed the San Vicente River, and struck off into the jungle. From that time until they struck the camp of Dagujob on June 4 they carefully avoided all trails and all open places. They picked their way silently through the bush, many times making wide detours to avoid clearings. The rate of march was very slow, averaging about two miles per hour.

Beginning with the departure from Catubig, Captain Stacey had made careful notes of the streams crossed and the direction of the march by means of compass bearings. For two days the troops marched in a terrible tropical rain that soaked clothing and provisions. Two-thirds of the ten days' ration of rice was rendered unfit for use and was abandoned.

Stacey permitted no noise while on the march or in camp. The men were not allowed to talk, and orders were given in whispers; tin cups and mess kits were filled with grass and packed carefully to prevent a rattle of sound. The camps were made in gulleys and in the heart of the forest, and all fires were fanned continually to prevent any smoke from disclosing their whereabouts. No wood was chopped; the fires were fed with small pieces or with large ends pushed carefully into the fire. So silent was the command and so successful the concealment efforts that on the evening of June 2 they pitched their camp in a hollow within 500 yards of three natives working in a field, without being noticed. The movement of more than 200 soldiers and *cargadores* in this manner was a tribute to the bush lore of Stacey. No lights or fires were allowed in the camp after dusk; Stacey cooled the ground under the fires with water before breaking camp, in order that investigating *pulajans* would be unable to determine whether the fires were new or old.

The camp was made on the night of June 3 within one and one-half hours' brisk march of the camp of Enrique Dagujob. At daylight the march was resumed, and Stacey took more than five hours to cover the distance. At one time he was so close to the fires of Captain Andres that the smoke was plainly seen through the bush. This camp was on the main trail to Dagujob's camp and about 600 yards to the north. Stacey made a wide detour through the bush in order to strike the main camp on the flank. They came in sight of the camp at five minutes after ten in the morning, the first view coming from the top of a small hill scarcely more than 100 yards distant.

Here, on the hilltop, Stacey left his *cargadores* and had his men remove their haversacks. He took stock of his situation. The evening before, his First Sergeant had injured his leg while in camp and was unable to march. Stacey had left him with seven men, entrenched on a hilltop near water, with three days' rations, in a place admirably suited for defense. This left Stacey, on the morning of the attack, with 71 men available for the assault.

Of this number the Captain detached 20 men under Sergeant Preston Ayres as a guard for the *cargadores,* and with twenty-two men of Company E and 29 Scouts under Captain Taylor, Stacey made provision for immediate attack.

The camp of Enrique Dagujob, *Jefe Principal de Pulajans de Samar y Leyte,* was situated in a small level valley entirely surrounded by an arroyo about 8 feet deep and 12 feet in width, and hemmed in on all sides by hills about 100 feet high. Stacey attacked from the west, with the sun directly ahead. He led his men down the hill, and they were within 40 feet of the most westerly houses of the camp before they were discovered. Stacey immediately saw that there were at least 400 fighting men in the camp, and he decided to kill the men in the houses nearest to the attacking force before attempting to cross the arroyo in the face of bolo attack.

He lined up ten men of Company E and opened fire. After a few rounds had been fired by each man, bayonets were fixed and the Captain charged across the arroyo at the head of his forty-eight men. In spite of all his carefully laid plans, Stacey had chosen an unfortunate approach to the outlaw camp. His line of attack was directly across the main latrine that Dagujob had established in the arroyo.

But at that moment, in 1905, Stacey was concerned with greater matters than the condition of his uniform. As he appeared over the edge of the arroyo the guide with him dropped, shot through the heart. A moment later a private fell, badly wounded. But then Taylor and his Scouts were across, and the force fought its way slowly through the camp. Taylor turned to the north after the camp was entered and engaged the force of Captain Andres, who had hurried from his outpost camp at the sound of the firing. Stacey, forty yards down the main street, was under direct fire from forty armed *pulajans* who were in the van of the attack. The *pulajans* were armed with Springfields, Krags, Remington rifles, Colt revolvers, and shotguns.

Thirty yards distant Commandante Felipe Senobio was rallying his bolomen for a charge and Stacey killed him with one shot. No stand was made by the *pulajans* after the death of Senobio. In the dogfight through the camp which followed and which lasted twenty-five minutes in all, every *pulajan* in the camp was killed, including Dagujob, Senobio, Captain Barnobal, Captain Titulado, and more than ninety *pulajan* soldiers.

In the camp Stacey found two prisoners who had been

undergoing torture. One of the men was lashed hands and feet on the ground, with his body arched away from a sharp bamboo stake that was set in the ground beneath the small of his back. Another was lashed in a standing position, bent forward over a sharp stake that would impale the victim in the stomach as soon as he collapsed from weariness.

When the battle was over Stacey found a strongly built *pulajan* lying on the ground near one of the houses. It was Dagujob, leader of the *pulajan* movement in Samar, and he was dying of several gunshot wounds. Stacey turned away and went to supervise the preparations for the return, leaving the *pulajan* leader in charge of a Sergeant. Some moments later the Sergeant came to report that Dagujob was dead, his end hastened by a severe bolo wound in the stomach. When pressed for an explanation, the Sergeant described the death of the *pulajan* leader.

A young camp servant in Stacey's party had come to stand before the dying *pulajan*. For a moment he had looked at Dagujob, and then he had spoken, "You do not remember me, Dagujob?"

The *pulajan* had shaken his head.

"I am Pedro, who was a servant in a house in a *barrio* you raided last year. It was then you cut off my ears and carried away my young sister into the jungle. You are slow in dying, Dagujob."

Then the youngster had lifted a bolo and stabbed Dagujob in the stomach.

Stacey reproved his Sergeant for allowing the servant to bolo a dying man and the Sergeant replied, "It was only a question of moments anyway, and seeing as this was a personal affair, I didn't see fit to interfere."

Stacey, remembering American soldiers he had found with their feet cut off, standing upright in wet sand where the *pulajans* had left them; remembering men with intestines nailed to tree trunks and then forced to run around and around the tree at the point of a bayonet until they were disembowled; remembering men in ant hills with honey on their faces and men impaled on sharp bamboo, was inclined to forget this little incident and not deal too harshly with a Sergeant who had conveniently looked the other way as Pedro had lifted his bolo. . . .

The men were issued rations in the center of that dreadful camp of dead *pulajans,* and as the sentries fired into the bush at the last of the retreating outlaws, their comrades in arms sat down to lunch surrounded by the bodies of the slain. The cook moved the bodies of the dead to make room for his fire, and there in those ghastly surroundings, with the customary tact of army cooks, served the men portions of canned roast beef.

Shortly after noon the troops departed for Catubig, carrying with them the wounded and the body of Dagujob. The dead outlaw was lashed to a bamboo pole to be carried back for identification. During the course of the long hot march, Dagujob became almost as offensive in death as he had been when alive, and Stacey was forced to relieve the men who carried the dead bandit at fifteen-minute intervals. The Filipino is not noted for any delicate sensibility to smells, but Dagujob so made his presence known during that march to Catubig that the carriers would complain to Stacey, "*Dios mio, Capitan, no puede aguantar . . . no puede aguantar.*"

In this manner Dagujob was returned to Catubig, and there, in the market place of the *barrio,* the villagers gathered to view the remains. The *Presidente* raised his hands, and three times he spoke to make the identification complete and official. "This is Dagujob . . . this is Dagujob . . . this is

Dagujob." And then they buried the *pulajan* leader and—in a measure—peace came to Samar Island.

But only in a measure, as we shall see.

Two months later, on August 16, Antonio Anugar, who had been the scourge of the Gandara Valley, was met by a patrol of Constabulary under Lieutenant George A. Helfert and Lieutenant Juan Sulse. Sulse killed Anugar in the course of a running fight in the deep forest, and his band was broken up and scattered. In this action the last of the rifles captured at Balangiga were recovered.

Under the heavy concentrations of Scouts, Constabulary and regulars, the resistance of the *pulajans* began to weaken. The island was systematically covered, with each branch of law enforcement keeping to its assigned territory.

The result of army operations and especially the death of Dagujob, had been to cause a breaking off of several thousands of natives in northern and eastern Samar, many of whom had been actively engaged in pulajanism and others who through fear had given support to the outlaws. Those who had presented themselves at the various stations, had been established in temporary camps until it was safe to return them to their former villages. The most needy were assisted, and those able to work were allowed passes to go out for food and hemp, which was their main article of trade. A strict reconcentration would have produced much suffering and would have brought about no good results at this stage of affairs. The end was not in sight, for the repulses in northern and eastern Samar had resulted in the transfer of *pulajan* operations to the southwest coast. There were some leaders and many of their followers who would not surrender as long as they could manage to move about from one part of the island to another. Nearly all of those men had been guilty of arson, torture, and murder. Preparations had been made to press the campaign regardless of the rainy season, with a view to forcing all organized bands to disperse. Once they were broken up, it became possible through the aid of friendly natives to locate individuals and guns, but the recovery of arms was possible only after patient and laborious work.

When the first outbreak occurred in the Gandara Valley, it appeared on the surface to have been purely the result of dissatisfaction of native hemp workers with the agents of the large commercial houses which practically controlled the market. This cause of dissatisfaction continued to exist as long as the *presidentes* were in control of the only form of government with which the poorer natives came in contact, particularly as long as those same *presidentes* continued to serve as business agents for firms whose sole interests lay in securing hemp at the lowest possible price and selling rice at a correspondingly exorbitant figure. There was no relief for this except with the gradual development of the Islands, which brought increased transportation facilities and, as a consequence, competition. These conditions were for years the cause of ill feeling amongst the natives, and a criminal element was always in waiting to take advantage of isolated situations and lead in disorders.

Public opinion as understood in the United States, which makes it odious to shield and conceal criminals, did not exist to any great extent in Samar.

Notwithstanding all this, it was felt that Samar was worth saving. It is an island of wonderful resources, and could easily absorb a great increase in population. There are many beautiful valleys which will produce rice and probably sugar, while the highest grade of hemp grows in all the foothills. The value of copra grown about the island is enormous. The

Catubig Valley and the many small valleys debouching in it, will furnish homes for a hundred thousand people from other islands whenever they become overcrowded. In fact, as early as 1906, it was suggested that a wise policy would be to establish settlements of natives in Samar to assist in its development and at the same time to help end disorder.

It seems proper, at this point, to mention the help given by the Navy to the various organized land forces that operated in Samar. The Admiral of the Fleet placed at the disposal of the Department Commander five gunboats and two launches, all under Lieutenant Commander Hugh Rodman, U. S. Navy. The moral effect of the presence of these vessels about the island of Samar was very great, and their services in moving troops to exposed points and in standing by several towns threatened with burning until troops could be brought in, had been the kind of assistance that counted for much. The cordial co-operation of the younger generation of the Army and Navy on this service laid a good foundation for harmony and success in future operations.

But there was still a sputter of resistance, and several more bloody hand-to-hand fights were to result from *pulajan* ambuscades.

On July 28, 1905, occurred the first of these affairs. Captain Ralph Jones, Captain William Green, Lieutenant Hemmett, and sixty-eight Constabulary were attacked by a large force of *pulajans* as they traversed a narrow trail. The mountaineers swarmed upon the police from both sides of the trail and were able to get hand to hand before they could be stopped by rifle fire. Six of the Constabulary were killed, and Jones and five of the men were wounded. The Constabulary withdrew in good order after inflicting heavy casualties upon the *pulajans*.

In March of the next year, the Constabulary suffered the terrible disaster at Magtaon. Here occurred the greatest massacre of Constabulary in the history of the corps. To the Insular Police the word Magtaon brings up the same memories as are the portion of the army men when they remember Balangiga.

Today, this town of grim memory is known as Concord. It is a tiny town on the south coast of Samar. In 1906 it was the scene of great activity on the morning of March 24. George Curry, an old-time officer of Roosevelt's Rough Riders, had been appointed Governor of Samar. Curry was very familiar with the natives in his district; he spoke their language and he was in every respect a good administrator. For months he had been negotiating with Aguilar in an attempt to persuade the *pulajan* to come in and surrender. The Governor had ridden into the mountains to visit Aguilar in his camp, and the *pulajan* chief had repaid with several visits to the Constabulary post at Magtaon. At last all details were arranged, and Aguilar agreed to surrender. At the suggestion of Governor Curry, the men unloaded their rifles to show confidence in the *pulajans*. The affair was to be made very ceremonious.

Aguilar, with 130 men, marched into the station and halted in a line facing the Constabulary, and some forty yards distant. Captain Jones of the Constabulary was standing in the rear of the *pulajan* line with the dignitaries who were there to see the surrender. He was chatting with Judge Lobinger, Superintendent of Schools Hoover, and two government officials named Scott and Campbell. Captain Bowers and Lieutenant Puno were in front of the *cuartel* with their men.

As Jones talked with Judge Lobinger there was something in the manner in which the *pulajans* waited that aroused his suspicions. He turned and leisurely made his way to the *cuartel.* As he reached it and attempted to speak to Captain Bowers, Aguilar gave a signal. The *pulajans* fired a volley and rushed the Constabulary. Fourteen of the Constabulary became separated by the sudden, unexpected advance. The *pulajans* ploughed through the center of the police line with their long crescent knives. Jones, Bowers, and Puno stood their ground and rallied the surviving Constabulary as they cleared a path and shot holes in the pulajan ranks with revolvers. A spear whistled to pin Jones through the chest; he extracted it and remained on his feet. Bowers received a gunshot wound through the left arm, but he too remained in action.

The fighting lasted but a few minutes. As the Constabulary fell, their rifles were seized by the *pulajans,* who made off into the jungle. In less than five minutes only seven of the Constabulary were on their feet. Twenty-two were dead; seven were seriously wounded, some to die later, and all had suffered minor bolo cuts. The place was a shambles, and fifty-seven dead *pulajans* and Constabulary locked in last frenzied embraces within the post grounds.

As the fight had begun, Governor Curry and the other government officials had escaped to the river. The Constabulary lost fifteen rifles in the action. Seventeen Medals of Valor were awarded the participants in this heroic stand. With Jones, Bowers, and Puno, the soldiers Villas, Barboza, Cunanan, Castro, Llorando, Abobo, Macariola, Bobo, Cuello, Cipriano, Fumar, Lopinac, Nofes, and Tazon were honored. Barboza died of his wounds a month later.

Jungle Patrol

THE REGULARS IN SAMAR

Cromwell Stacey, transferred now to the Philippine Scouts with the rank of Major, had meanwhile been ordered by Major-General Wood to make a report of suggested recommendations for the best means of bringing order to Samar. This Stacey did, after conference with General Allen of the Constabulary.

In all particulars Major Stacey's suggestions were carried out, and in no small measure much of the success of the pacification measures on Samar Island may be credited to this first-class fighting man.

During the year between July 1, 1904, and June 30, 1905, the United States regulars, the Scouts, and the Constabulary were in almost constant conflict with the *pulajans*. The fighting reached a virtual end with Stacey's assault on Dagujob.

A fact not generally known by the casual reader of the history of the fighting in Samar is that the *pulajans* were organized into a complete military corps. The hillmen were organized into regiments and brigades with Line and Staff officers. In the Dagujob campaign, Captain Stacey brought to light the interesting fact that in the *pulajan* army, as in more civilized armies, there was continual bad blood and argument between line and staff. Included in the mass of *pulajan* correspondence captured by Stacey was a complete file of the letters that had passed between Line and Staff, the Line officers complaining bitterly that their troops were unable to march on the trails due to army, Scout, and Constabulary, and the Staff countering with the suggestion that the troops operate only at night.

The outcome of this correspondence seems to have been that the Line adopted the advice of the Staff, for some weeks later while Stacey was in a jungle camp, his outpost sentries reported a long line of flickering lights which marked the approach of a column of *pulajans* moving in accordance with Staff instructions. The troops under Captain Stacey held their position in perfect silence until the *pulajan* column was within fifty feet of the sentries, and then opened with a volley of rapid rifle fire. In the next batch of correspondence between the *pulajan* Line and Staff, the incident was reported by the Line officers with great indignation.

Major Hugh D. Wise of the Philippine Scouts contributed another potent source of friction within the *pulajan* ranks. This Major, while on an extended patrol in *pulajan* country, sought shelter from a tropical rainstorm in a native hut. Happening to glance up to the roof tree of the hut the Major noticed a red bandanna handkerchief suspended from the roof beams. Curious, the Major scaled to the roof and found the handkerchief to be filled with .45 Springfield shells, ready for delivery to the *pulajan* ordnance section.

While waiting out the rain the Major carefully removed the powder from all of the shells and replaced it with dynamite. Sometime later the *pulajan* official correspondence became filled with fierce letters between Staff and Line. The Line complained bitterly that so many men had been killed by defective shells that the men were afraid to fire their rifles, the Staff countering tartly with the suggestion that the line see that the men kept their arms clean and in serviceable condition.

All of which was obviously to be traced back to an idle afternoon in the rain when Major Wise had filled some dozens of *pulajan* shells with dynamite. Official reports of the day mention the finding of dead *pulajans* lying beside their shattered rifles.

Meanwhile, this quality of resistance had indirectly aided the hard-pressed Constabulary forces in Samar. In April, 1905, authorization was received for the purchase of 500

Krag rifles with bayonets. Three hundred of these weapons were issued to the Constabulary by the middle of the year.

There were still many deficiencies in the organization of the Insular Police. The ammunition issued was old, and the shells were prone to stick in the rifles. But with the gradual arming with Krags, the odds became more even in that terrific struggle for the mastery of Samar Island.

Chapter xiii

MOHAMMEDANS

"Officers and enlisted men embarking in small boats will lash rifles, revolvers and heavy equipment to the boats to prevent loss in case of capsizing."

—*Constabulary Manual*

IN the south was the great shaggy island of Mindanao.

That mystery island that shades the equator had been the scene of the greatest conflicts of all, long before the birth of the Constabulary. That island had seen Malay against Ming; Moro against *conquistadore;* and brown men against brown. It has been a battleground for many breeds of men who have scrambled for a foothold on its soil.

There is a feeling of permanence about Mindanao that seems to hang in the air. One has but to step upon its gritty sand beaches and look back along the rolling, jungled hills to know that here is a land which is stern in resistance; that here is a land which is pregnant with unpleasant memories and bristling with unwritten stories. It had been stained with the blood of a dozen races of men.

It was a land where illusions and men had died—and where more men were to die.

Mindanao has no history: it *is* history. It was pioneered by the Portuguese, who received no historical credit. Casual navigators, these, who landed upon its coast a decade before the official "discovery." Men followed after them, as men had gone before. Spaniards with Toledo blades and glittering armor; red-coated Englishmen, and Frenchmen with waxed *mustachios*. Not to mention Chinese, Javanese, Japanese, and other mixed breeds of the Orient. Jesuit priests had hoisted a cassock to do battle with the Moros as they had

A. E. Hendryz

Leonard Furlong

Bud Dajo — Scene of Greatest Battle of the Moro Campaigns

American Troops at Camp Kiethley in Lanao Province

sought, unsuccessfully, to replace Koran with Catholicism . . . after them, Americans in blue shirts and slouched campaign hats.

It had been a pageant of history and races that had passed before Mindanao—gaudy names had struggled there—but not a King of all the world had been able to say with honesty, "This island I own."

The year 1906. . . .

Commanding officer's quarters at Pettit Barracks, Zamboanga.

General Leonard Wood penning a note to Colonel J. W. Duncan of the American assault forces. "I wish you would get two of your companies together and go to Jolo at once. Nothing but blanket rolls, field mess outfit, seven days' field rations and two hundred rounds per man. In haste. Regular orders will reach you later."

The storming of the *cottas* on Bud Dajo . . .

Up the steep sides of the mountain, American soldiers advance against the frowning *cottas,* garrisoned with 1,000 armed Moros. Nine hundred and ninety-four Moros fall before the Gatling guns of the twentieth century. The trenches are piled high with the slain. Kris against Gatling gun. Spear against mountain artillery.

Civilization marches on. . . .

The year 1933. . . .

Malacanan Palace, Manila. American officials bend over a bulletin, confronted again with an age-old Moro problem. "Third Lieutenant Mariano G. Esculto, commanding officer of Camp Andres, was killed when a patrol he was leading was ambushed by Moros at *sitio* Cambusi, near Camp Andres, this afternoon." (*Manila Daily Bulletin* for November 21, 1933.)

An American Governor-General stands by—in the same hall where de Sande and Figuroa stood so long ago. He confers with grave-faced men in khaki as they read on in an official bulletin: "The report of the bloody encounter, which reached here late this evening, threw Jolo into considerable excitement as this province has been quiet since the killing of Lieutenant Julio F. Barbajera on *September sixth*. People here thought that the government had the Moro situation well in hand, following the conferences held in Manila between Governor-General Frank Murphy and Lieutenant-Colonel Luther R. Stevens, District Commander of Mindanao and Sulu."

December 1, 1937. . . .

The Associated Press wires carry a footnote to the conquest of Mindanao, a footnote unfinished by Portuguese and Spaniard and Dutchman and Englishman. The press despatch reads: "Three powerful planes took off today from Manila, bound for distant Mindanao Island, as the Philippine Army launched the most determined campaign in the history of the Commonwealth to crush age-old banditry in Lanao Province. Simultaneously, 116 infantrymen and officers sailed for Lanao aboard the inter-island steamer *Samar* to join other Insular troops who have conducted a week-long siege in an unsuccessful attempt to dislodge Moros from a half-dozen *cottas*. Departure of the bombers marked the first time in the history of the Philippines that aerial weapons have been used in the Moro campaigns. The planes, one of which bears Major General Paulino Santos, Army Chief of

Staff, will be used to attack the *cottas* in the event trench mortars are unsuccessful.

"One soldier was killed and nine wounded last week when Insular troops attempted to storm the outlaw stronghold. The *cottas* have been tunneled into the side of the hill and so far have proved impregnable to infantry attack."

This then, was the land that America had "bought" from Spain, but there was to be more to the purchase price than the $20,000,000 she had paid. America, as others had, was to find death in Mindanao.

The purveyors of that death were the Moros, and no book would be large enough to record the history of these Mohammedans who have so smeared the pages of Far Eastern history with blood. No book even remotely concerned with the pacification of the Philippines could ignore the Moros. As the races of men ebbed and flowed across the face of Mindanao, like a disordered tide, the Moro raised a kris blade against them all. Joyfully, he raised that blade. The savage resistance of the Mohammedans to the *conquistadores* of Spain must stand as the most amazing epoch of military history—377 years of uninterrupted combat!

The Constabulary made a tentative gesture toward the Mohammedan country in 1903, but for several years their activities were to be secondary to the regular army. For almost a decade Mindanao and Sulu were to be the problem of Generals Leonard Wood, Tasker Bliss, and John J. Pershing.

Captain Pratt of the regulars had relieved the battered Spaniards in May, 1899, but the region was lightly held for several years. It was but a gesture of sovereignty at that early date. The future pacification was the result of the combined efforts of army, Scouts, and Constabulary, but to the army must go greatest credit for the initial establishment of American authority.

MOHAMMEDANS

When we purchased the Moro country from Spain as a part of the Philippine Archipelago, we contracted for its pacification—a matter Spain had never been able to accomplish. The Moro country was geographically a portion of the Philippine Islands; actually, it was a country to itself.

The Moros are a race who diet on blood and steel. They were ill-disposed to accept the authority of the United States. They had never heard of the Treaty of Paris. Nor had they heard of a nation called the United States of America. Almost from the beginning they fortified their bamboo forts against encroachment. It was a matter of routine for them; they had been doing that for years. We were a new set of strange faces—that was all.

American soldiers began to learn of *Imams,* those white-robed priests of Mohammed, who are stern in purpose and unwavering in faith.

Behind the *Imams* were their official agents of death—an iron ring of swarth, corded krismen; that iron tempered into chilled steel by the dictates of the Prophet. A people who thrive where white men die. A race galling in battle, and a proud people who seek welcome death on the field of battle.

Even today, in Mindanao, with the combat noises muted, the people seem always poised for battle.

The Constabulary in this country was officially organized in 1902, but it was not until early in 1903 that Captain (now Major-General, retired) James G. Harbord was assigned to command the district from headquarters at Zamboanga.

To form the newly organized force, selected noncommissioned officers were detached from the companies of the northern provinces to form a nucleus of veterans. The work of recruiting Moros to fight Moros was undertaken speedily. The decision to attempt the recruiting of Moros into the service of the United States is probably the most splendidly audacious move in the entire history of the Constabulary. No one unfamiliar with Moros and Mindanao could appreciate the qualities of officership required to make this organization successful.

By the end of 1903, the force consisted of 17 officers and 353 men (Filipino and Moro, and in itself a marvelous disciplinary feat), and they were on post at Surigao, Cagayan, Oroquieto, Zamboanga, Tucuran, Siassi, Bongao, Baganga, and Mati.

Immediately Captain Harbord encountered peculiar difficulties incident to the religious beliefs of the Mohammedans. The uniform was modified to the convenience of the Koran; as the Moros would not wear hats with brims, they were issued red fezzes with black tassels. So natty was this headgear that many of the officers adopted it as a part of their uniform.

In the vicinity of Zamboanga, the Constabulary was composed of equal portions of Moros and Christian Filipinos. It was necessary to establish separate messes in each company, as the Moros were forbidden by the Koran to eat pork. This ban prohibited the old army stand-by of pork and beans, as well as the issue of hardtack, which contained lard as an ingredient.

The Moros came slowly to the Constabulary. By the end of 1903, one hundred and two of the Mohammedans had been enlisted.

The greatest problem of the authorities was that question of the feasibility of arming these wild Mohammedans with rifles. The Moro youth purchases his bride, or brides, from the father-in-law, the desirability of the maidens influencing the purchase price. It was pointed out that a rifle was an object almost worth its weight in gold in the interior mountains, and that many young Moros might enlist with the idea of prompt desertion with their arms.

But the Moro has a peculiar, inflexible code; once he became accustomed to military discipline, he proved a loyal soldier, and desertions were almost zero in the force after organization was complete. Captain Harbord deserves great credit for the tact with which he explored this dangerous possibility.

Harbord made an effort to undertand Moros, and he learned many things about these strange grim men. He discovered that to change officers too often was to affect discipline. The Moros made a personality out of their officer; if he was brave and fair, they idolized him for those two prime virtues. Officers had to be especially strong in those qualities of leadership and valor.

Behind that type of officer, the Moros would go smiling and happy to their deaths.

The Moro was a soldier, no question of that; and when sworn into the service of the United States his stern nature made duty an inflexible pleasure. On one occasion, when Lieutenant Wood was Acting Chief Justice of the Tribal Ward courts, he had in the guardhouse a Moro prisoner convicted of murder and sentenced to be hanged.

Wood felt badly about the hanging, as the victim was a brother of one of his most trusted Constabulary soldiers. On the eve of the hanging the Lieutenant visited the guardhouse and found to his surprise that the Sergeant of the Guard had detailed the soldier to guard his jail-bird brother. Naturally Wood wondered if conniving for escape was the plan. He decided to leave the guard as it was, and retired to his room, from which place he could oversee the situation.

When dusk settled and the sentry was at the farthest end

of his beat, the prisoner burst from jail and ran for the wall. His brother, the sentry, turned and saw him. He paused for a moment, and then laid his rifle on the ground. In a flash he was racing after his escaping brother.

As Wood watched, the sentry overhauled the escaping prisoner. There was the flash of a bayonet in the dim light —and no necessity for a hanging in the morning!

The Constabulary found an entirely different situation in the Moro country from the conditions that had prevailed in the northern islands. There, in Mindanao and Sulu, the disorders were not temporary post-war disturbances or incidental to the era of insurrection, but were, and for centuries had been, the normal order of affairs.

No government had broken down in Mindanao. No government had existed there! Two Sultans and a great number of *Datus* had controlled the actions of their personal followers. The Sultan at Jolo was supreme in matters pertaining to the religion; there his authority ended.

The stern Lawarn Code and the Koran were the law, but the kris overshadowed the Koran as an instrument of justice. The problem confronting the government was how to stop this "merry carnival of human sacrifice, murder, slavery, kidnapping, cattle-raiding and piracy," and to convert the inhabitants of this vast jungle into useful citizens capable of a certain measure of self-government.

Young Constabulary officers of selected character played a great part in the civil affairs of the country. Many of them were appointed Acting Chief Justices of the Tribal Courts, serving as patrol leaders in capturing bandits and then as judges of the court that condemned the bandits. Quite often, unusual judicial conditions prevailed.

MOHAMMEDANS

Lieutenant James L. Wood was one of these jungle justices —a traveling justice similar to our Circuit Court justices. His position also as a Constabulary Lieutenant often placed him in a peculiar situation. Naturally, it was not in accord with judicial dignity that the Judge should wear a pistol, but the fact remained that the Judge (in his capacity as a Constabulary officer) was the object of concerted Moro attack, even in the Court chambers.

While Wood was serving as a Judge, General Bliss realized the position of the Lieutenant and sent him a shoulder holster that bore an interesting history. It was constructed so that a small, leather-covered piano wire fitted into the muzzle of the revolver, with a hand-spring circling the cylinder. When wanted, the gun would fairly leap into the hand.

The holster had been given to General Bliss by his lifelong friend, "Whispering Smith," who had designed it and carried it for many years for the Santa Fe railroad, and who was rated as one of the fastest gun men of the early West.

The holster had been of particular interest to Wood, for his initial experience in Mindanao had almost been his last, due to a stiff holster. He had been a Third Lieutenant then, only three days in Mindanao, and had gone with a group of brother officers to the wedding of a friendly Moro. The other officers had worn no side arms, but Wood a youngster of twenty-one, had been proud of his new service revolver and holster, and had worn them. The officers were given a place of honor on a high platform.

The groom approached, with bales of silk, brass gongs, and a retinue of shackled slaves to complete the purchase of the bride. As the wedding party watched, one of the slaves escaped into the bush, and the young Moro was disconcerted and lacking in the payment. Hastily he asserted the prerogative of his rank and clapped the nearest of his men into the slave line. It was a matter that could be straightened out later, for this policy of degrading a free tribesman was contrary to Moro law.

The procession moved on, the young free Moro sullen in the slave line. The wedding got under way, and the slaves were seated on the floor in front of the officers' platform. As interest turned to the bride, the young free Moro seized a barong and went to work on the crowd. The slave keeper dropped, split from shoulder to pelvis; the Moro turned to the officers' platform.

Wood alone was armed—and his holster was stiff and new. For the longest seconds of his life he tugged at the weapon—it came free at last and the Moro went down under Wood's accurate fire.

That evening, a solemn-faced young Constabulary Lieutenant went to his quarters to accomplish an exemplary job of holster pruning.

So Wood, a few years later, was grateful indeed to sit on the judge's bench with "Whispering Smith's" holster beneath his arm.

The Moro Province was under a special form of semi-military rule, with a Commander of the United States Army acting as Governor. It was desired, as soon as was possible, to convert this military régime into a civil government. With that in mind, the Constabulary was hurried into organization to be the agents of that government.

Between the years 1903 and 1937 the Constabulary was to engage in hundreds of *cotta* fights, and to quell twenty-six major uprisings of sufficient seriousness to be listed as "campaigns."

Harbord was indeed busy in Mindanao and Sulu.

He had whipped the Moro Constabulary into condition to take the field—in itself a most remarkable feat. The Moro is one of the finest individual fighters of any race of men, but he is not amenable to military discipline. It required a man of great quality of leadership to accomplish that result: Harbord had proven himself the man for the job.

No one who ever knew General Harbord could forget that peculiar walk with the little twitch of the left shoulder. Even the Moros remembered it. When Harbord first came to Mindanao, in 1903, he was already slightly bald, disguising the red hair of his youth. He has changed very little with the years.

There was certainly an element of splendid organization and sympathetic understanding in the nature of Harbord to enable him so to win the confidence of the savage Mohammedans. They were ready to fight in 1904.

And there was fighting to be done. . . .

A line of *cottas* was stretched across Mindanao, and Sulu and the Moros were massing to resist the American invasion of their country. They had some 34,000 official warriors in the field, but every man capable of bearing a kris was dangerous. The total Moro population of the southern islands was about 400,000, and the fighting men were distributed with 19,000, in Mindanao, 10,000 in Sulu and about 5,000 on Basilan Island.

The Moro *cottas* were forts of bamboo and *nipa*, with stout double walls of tree trunks packed with earth or broken coral. They were defended by *lantakas,* which are ornately carved, swivel cannon of the Moros, and are older than the written history of the archipelago. Some authorities have stated that these cannon of the Malay pirates were among the earliest form of portable ordnance. They fired a ball weighing from one quarter of a pound to twice that weight, and they were adaptable for use on land or mounted in the bows of their vessels. The pirate *garays*—the long, outrigger boats with bulging lateen sail—used the *lantaka* on their slave-stealing expeditions through the East Indian Archipelago.

But the most stern line of defense of the *cottas* was the line of individual combatants who waved the great fluted and wavy-edged blades that are a marvel of steel craftsmanship. Elsewhere, it has been written that these Moro krises were the equal, in temper and edge, of the best blades of Toledo and Damascus.[1]

In 1903, the Constabulary was not in sufficient strength to consider the subjugation of the Moros, nor did they reach that state for some years. In the beginning, the conquest of Mindanao, and Sulu was strictly an army job, and the regulars were in the field officially against the Moros until the battle of Bud Bagsak on June 16, 1913.

The regular army took the field with magazine rifles, with Vickers-Maxim machine guns, with mountain artillery, and with the finest officers of the military establishment in command. Generals Pershing, Wood, Bliss, and countless others acquired their military reputations fighting against the Moros in the Philippines.

There can be no question but that the regular army of the United States was very efficient in the Moro country in that ten-year period of combat with the Mohammedans. The decade of resistance the Moros set up against our regular army is a most remarkable feat of arms. The long-range tactics and the rapid-fire weapons of the army, with few exceptions, prevented the Moros from getting to close quarters. The result was an unfortunate period of unequal warfare that aroused great criticism in the United States.

1 The reader interested in the Moro, his weapons, mode of attack, customs, religion, and early history, is directed to the author's *Swish of the Kris* – their military history. This account carries the Moros through four centuries of their fighting existence.

MOHAMMEDANS

The late John Hackett, for many years editor of the *Mindanao Herald* of Zamboanga, and in the early days a civilian information operative working with the army, estimated that 15,000 Moros met their deaths in consequence of the campaigns. This was accomplished with a casualty list in the regular army that was virtually negligible.

In considering the argument from the military viewpoint, there can be no question but what something had to be done about Mindanao and Sulu. Murder and piracy had been unrestrained there for centuries; the lives or deaths of men were held in the palms of the Moro *Datus* who were the law of the land. Such had always been the rule in Mindanao, and the Moros had seen no reason for a change because of a title change in land ownership due to our purchase of the country from Spain.

These Mohammedans are not pretty objects, nor are they sympathetic objects, but the warfare they waged against the United States was justified. They had resisted the entrance of a dozen races of men; there was no valid reason for their excepting American troopers.

And they did resist—with the most severe resistance that our arms have encountered in the history of our nation.

In 1903, the regulars under Leonard Wood had a series of serious *cotta* fights in Lanao Province in Mindanao while Major Scott was engaged in a lengthy campaign after the *Datu* Hassan in Sulu. Hassan had sent word to the American government to the effect that "if they wanted him, they could come and get him."

Scott got him—after many weeks of weary warfare.

During the next year, the campaigns after "Ali in the Valley" occupied the attention of the army. Ali had organized the Moros of Mindanao into a tremendous coalition. He was eventually surrounded in his *cotta*, completely surprised, and destroyed, with most of his family, by long-range bombardment.

The Constabulary had no part in these operations save for a minor collaboration with the army in a few inconsequential skirmishes with Moros. While the army campaigns were in process, the Constabulary was gradually extending its influence to the more isolated regions. The post at Siasi was taken over from the regulars on June 20, 1904. Siasi is in the heart of the Moro country on the island of that name. A few months later, the Insular Police took over the island of Tawi-Tawi with a station at Bongao.

On Basilian Island, a station was established at Isabela. Here, Captain Sandford, with 49 men, occupied the old Spanish fort that was a relic of the unsuccessful days of the *conquistadores*. On clear days Sandford could almost look into the bastions of the old fortress of Senora del Pilar at Zamboanga, fifteen miles away.

In Cotabato Province, 95 men under Captain Long were occupying stations at Cotabato, Libungan, Tumao, and Taviran. To this sector came two raw young officers who were extremes in temperament, nationality, and appearance. Lieutenant Gilheuser was one, a big blond German, fresh from the Prussian army.

The other was a slim, dark American boy, very carefully groomed and sporting a tiny mustache. As one considered his slight frame and slender wrists, he seemed ill-adapted for service in this organization of rough fighting men who took jungle in their stride.

But he was to develop—in the opinion of many—into the greatest warrior of them all. His name was Leonard Furlong. His short, wild life must have been an unhappy one. There could have been no other reason for that dashing, frenzied career of battle that he waged in Mindanao.

Twenty years after Furlong had fired his last shot, this writer stood with wrinkled and ancient Moros on the sites of some of the Cotabato battles of this Captain of Constabulary. We talked, the Moros and I, of those old days of murder and piracy and ambush, when the kris had been the law and the measure of a man. The Moros are always ready to talk of battle.

These scarred old reprobates with blackened teeth and betel-stained lips, were no exception. Our conversation that day was filled with grand names: Allan Fletcher of the Scouts, called "Papa" by Moro and Filipino and American—a grand campaigner; Lieutenant Whitney of the prodigious strength gained a shuffle of bare feet and the twitch of a turban; then we talked of a Lieutenant named Cochrun—"a brave man, *si,*" was his accolade; a youngster's name came into the conversation—Jesse Tiffany. The Moros fought him on their *cotta* walls. He, too, was valiant— a nod of the turbans confirmed him with the greatest praise a Moro can bestow on a man.

But when I mentioned Furlong, a glisten came into the eyes of ancient Moros who talk of redder and grander days. They sent up the most impressive salute to Valhalla that I can ever hope to witness. I see them now as I write—a circle of genial old ruffians, almost ready themselves to mount a white horse to Paradise. Their turbans are off now and their chins at rest on their scarred and brawny chests. After twenty years, they bend a neck to the memory of Leonard Furlong—"most desperate fighting man of all."

Furlong in Mindanao, Crockett in Samar: the Constabulary may well be proud of them.

Far away on the south coast, in Davao Province, the Constabulary had stations at Davao, Mati, and Baganga, with

Williams, Taulbee, Bernal, and Fort in command of a garrison of 115 men. The bloody region of Lanao was tentatively occupied with stations at Iligan and Misamis under the leadership of Griffiths, Wood, and Heartt. The dread region near Lake Lanao which was to be the scene of the bloodiest battles of the Moro wars was not under occupation. Misamis Province was in command of Captain Gallant, with Lieutenant Campbell and a police force of 78 men.

At Siasi, DeWitt and Sowers held forth with 40 men. Lieutenant Johnson was at Bongao with his 33 native infantrymen. Another Johnson was in Zamboanga, and a Moro in the mountains was sharpening the spear that was to skewer him on Basilan Island.

Surigao will be remembered as the station where the surprise attack had cost the life of Captain Clark. It was garrisoned in 1904 with 73 men under Captain Waloe and Lieutenants Lattimore and Burrell.

Zamboanga became headquarters for Constabulary and army operations in the south. For more than four centuries Zamboanga has been the headquarters of men who scrambled for a foothold in the Moro country. The town broods on the memory of a martial past.

As the years hurried by, Bagumbayan, the pirate village that was, became Zamboanga, the metropolis of Mindanao. The undertone of its growth was the mutter of battle, for the city has been little else than a military reservation and concentration point. There is no hinterland to promote a sound commercial growth, for the peninsula of Zamboanga is spined with a backbone of rugged mountains. Today, a few white prospectors have scaled those peaks in search of gold. Otherwise, they remain as God created them.

No visitor to the southern islands can remain long unaware of the brooding presence of the old forts of Spain. They were abodes of horror in the old days, and they seem to retain those memories of retching death on the kris. These great thick walls and towering bastions were a gesture of futility—a gesture of the Spanish occupation of Moroland.

The Spanish fought a waiting war against the Moros. Their indecision developed into the most lengthy wait of history. For almost four centuries they cowered behind walls of stone. The Spaniards took up the Moro wars with a great confidence and a brave flutter of pennons. They dropped the conquest with a feeling of apathy and stark fear.

Only in Mindanao, of all the world, did the Spaniard of the looting age fail to make conquest. There, the *conquistadores* met men who considered warfare one of the major pleasures of an otherwise drab existence. They met Malays who struck with ferocity and withdrew silently through the tall grass. They met men who were inspired by a major defeat, and who deliberately prolonged the warfare for the sheer joy of fighting.

Under the impact, the Spaniards wavered and sought walls of stone. Their occupation of Mindanao became mole-like. Their greatest engineering feat was the construction of the noble fortress of Nuetra Senora del Pilar at Zamboanga.

Today, this moss-grown old fort dominates the town of Zamboanga. Today, as in 1904, it is the headquarters of the regular army in Mindanao and Sulu. Tomorrow, it will be headquarters for the southern division of the Philippine Commonwealth army—and they will still be fighting the Moros.

With Zamboanga's old Spanish fort as supply base, the regulars began to spread out across the face of Mindanao and

Sulu. Their job was to accomplish what Spain had been unable to do. Massed men and rapid rifle fire replaced the arquebusses and the gaudy "Regiemento Rey y Riena" of Spain. America took to the open field. The old Spanish forts became supply stations. Vigorously, the army carried the war to the Moros.

The reverses they suffered were minor; they had too many guns and too much artillery. The Moros were never able to get hand to hand, except by surprise night attacks.

The Moros prowled the flanks like leopards, harrying the camps at night, striking quickly to kill a straggler, but were never dangerous except to small, isolated patrols. The army kept together, in large bodies of well-armed men.

In Mindanao and Sulu, exactly the same condition was to prevail as had been the case in the northern islands. The army blasted the Moros from their main positions and broke up the organized resistance. Then, as in the north, the army withdrew, and the Constabulary was left with the responsibility of the actual pacification of the country. The army campaigns were the usual large troop movement operations; the Constabulary operated with the small patrol system, in desperate hand-to-hand encounters.

The patrols were moving out, in 1904, to establish their chain of small patrol stations. Far across the straits on Palawan Island, the Constabulary set up four stations. At Cuyo, Coron, and Puerta Princesa, sergeants had charge of a few men on outpost duty. At Balabac, at the extreme southern tip of Palawan, Lieutenant Walker was experiencing the ultimate in isolation with his post of 45 men.

Palawan Island was a no-man's land, of no particular interest to American or Moro. The post at Balabac was a scene of grim memories. In the earliest Spanish days it had been an outpost of the *conquistadores,* and there 122 men of a garrison of 375 had died in a single year. The Spanish soldiers had so dreaded Balabac that they had suffered self-inflicted wounds to avoid station there. The malaria fever was malignant. Even today, Balabac is a forbidding place. To Lieutenant Walker, in 1904, it must have been a place where one lived on the jungle. It was a long sail, and reliefs were infrequent.

For more than three decades the Constabulary was to be battered by Moros in this southern archipelago of the Philippines.

Chapter xiv

YEARS SHOT WITH FLAME: 1906-1907

"As the primary object of a patrol is usually to gain information, its members will be instructed to make independent observation as the country is traversed, conversing freely with the inhabitants of the *barrios* concerning local affairs. Where a patrol is not in active pursuit of ladrones it is better not to traverse the country rapidly . . ."

—*Constabulary Manual*

AS the months succeeded each other, and the calendar reached into the year 1906, the Philippine Constabulary began to acquire a firmer grip on the northern islands of the Philippines. Only an occasional disturbance marred the serenity of the north. In April the old *Santa Iglesia* came to life under the leadership of Manuel Garcia. The Constabulary station at Malolos was battered by a wave of fanatics, and twenty-two rifles were lost, with 1,800 rounds of ammunition. All of the police force but three were killed or wounded.

Lieutenant Jose de los Reyes went in pursuit of Garcia, nineteen men supported by twenty men under Lieutenant Walker. They came to grips with Garcia at dawn in a misty rice field on the ninth of July, and in the engagement that ensued Garcia and most of his men were killed.

In Tayabas a religious-military organization called the *Ejercito Libertador Nacional* flared briefly with a burst of murder and robbery. Pantaleon Villafuerte set himself up as the defender of the expiring Filipino liberty. A young Filipino lawyer named Manuel Quezon, who was Governor of Tayabas, took the field with a Constabulary force commanded by Major Borseth. On July 11, Villafuerte was killed and nineteen of his officers surrendered. The *Ejercito Libertador Nacional* movement fizzed and went out. This secret society had as its aim a complete independence for the Philippines, and a pretentious military organization had been built which expected arms shipments from Japan. Two Lieutenant Generals were captured by Borseth and Quezon, a "Vi-Rey" and a Commandante General. Included in the batch of prisoners was one Lieutenant . . . possibly the only Lieutenant in the organization.

On May 20, 1906, the gallant old fighting Colonel, Wallace Taylor, was in the field with two officers and fifty-five men in the Province of Jautna, on Samar. Days of strenuous march had brought him into contact with Otoy, and the fortress of the *pulajans* was located and attacked after days of search in the deep forest. As the Constabulary moved in beneath the walls, Taylor was struck in the neck and jaw by a .45-caliber bullet that carried away a portion of the jawbone. Although greatly weakened by the shock and the loss of blood, the Colonel remained actively in the fight until the *pulajan* fortress was taken and destroyed.

It was five days before Taylor could receive proper medical attention, and he was kept alive by feeding him eggs found in the nests of wild chickens. He was awarded the Medal of Valor by a General Order dated September 21, 1906.

On Leyte Island the United States regulars were engaged in the final months of active operation against the small detachments of hill guerrillas. Basilio Sampson led a *pulajan* force against Lieutenant James of the 8th U. S. Infantry on August 9, and in the indecisive engagement that followed Lieutenants Treadwell and James were killed. The *pulajans* closed the year with an attack on Captain Ham, also of the 8th Infantry, and a detachment of twenty-five Constabulary under Lieutenant Yates. A bloody fight at La Paz resulted in the death of Sampson and forty of his men. Five American soldiers were killed and Lieutenant Yates was wounded.

Leaderless, the *pulajans* withdrew to a stronghold deep in the mountains where they were attacked by Captain Chrisman of the 16th Infantry and Lieutenant Hemmett with a Constabulary force. Hemmet was wounded in the assault which carried the *pulajan* fortress. This action at Dagami quieted the island of Leyte.

Across the channel on Samar, the back of the *pulajan* resistance was breaking. The *pulajans* no longer stood to fight. The warfare developed into a tedious man hunt conducted by small detachments of Constabulary. The year 1906 saw the extermination of the last of the hill chiefs. De La Cruz was killed in a Constabulary operation in November of that year. Major Murphy of the Constabulary killed "Papa" Pablo during the same month. Picardel was killed by Lieutenant Edmondson in January, 1907.

No longer did the red-shirted *pulajans* descend in force from the mountains. The last raid to the lowlands occurred in November, 1906. The opening of the year 1907 saw Otoy, Faustino, and Felipe Salvador at large. They alone, of all the great mountain chiefs, had survived the campaigns of that terrible year on Samar Island. Otoy, the genius who had inspired the massacre at Mugtaon, was killed by Lieutenant Puno of the Constabulary, who had been one of the survivors of the Mugtaon affair. Faustino was killed on June 11 by a detachment of regulars of the 8th Infantry under the command of Lieutenant Jones. Felipe Salvador remained at large, but inactive, until 1911.

The official reports of the Chief of Constabulary show the year 1907 to have been the turning point in the northern and middle island campaigns. Pulajanism was crushed. During that year, too, the complaints against the Constabulary ceased, and discipline and equipment was greatly improved. There were but twenty desertions during the year in a force of 4,748 men and 305 officers. The Insular Police were maintaining 167 stations, and were no longer in collaboration with Scouts and regular army.

One hundred and five thousand pesos had been appropriated for the purchase of Krag rifles, equipped with bayonet, and of six-shot capacity. Of these, 2,120 had been issued to the Constabulary.

The older officers were beginning to give way now. General Allen returned to his regiment, to be relieved as Chief by H. H. Bandholtz. Scott and Baker were relieved, to be replaced with W. C. Rivers and M. L. Hershey.

The scene of action changed as peace came to the north. All of the resources of the Constabulary were turned to the south, to Mindanao and Sulu, where the Moro wars were but beginning.

The initial subjugation of the Moros had been undertaken by United States regulars, who had waged desultory and not too effective warfare against the Mohammedans since 1899. Here again the large-scale operations of the army were not effective, as the army posts controlled only the immediate area surrounding the station. The regulars had been uniformly successful against the Moros, who had chopped viciously at the army flanks and dissolved into the jungle. The result was a great, unpacified jungle, filled with small roving bands of malcontent Moros, and completely out of the reach of the army brigades.

True, Captain McCoy had exterminated the *Datu* Ali in a bitter and bloody affair in Mindanao, and Pershing was well on his way to his ultimate post as commander of the army in France, through a series of operations in interior

Lanao. But in the main, Mindanao and Sulu had yet to feel the hand of America.

As the *pulajan* campaigns in Samar had faltered to a close, the Constabulary detachments in Mindanao and Sulu had been gradually strengthened, and a new fighting force began to make itself felt in the Moro country. New faces appeared, reckless youngsters who began to lead small combat patrols deep into the heart of Mindanao. Mass troop movements gave away then to that more primitive hand-to-hand combat which was the only solution to the problem of the Moros.

As the Constabulary patrols began to penetrate deeper into unexplored country, the temper of the Moros became more surly. A brand of fighting developed that for sheer ferocity surpassed the best efforts of the bloody days in Samar. The war was carried to the Moros, and right nobly they met the challenge of the Constabulary patrols.

In that desperate struggle for their home soil the Mohammedans won the respect of every officer of Constabulary who faced them. They were a valiant, minor group of disorganized tribesmen who found themselves in the path of the steam roller of conquest. They offered the most stern quality of resistance that America has encountered in our military history. They were defeated by schoolboys who combined tact and bayonets to accomplish what Spain had failed to accomplish.

It was not until the year 1906 that the Philippine Constabulary was a potent civilizing force in the Moro country. They arrived in time to be the spearhead of a terrible decade of jungle war. By February, 1906, the greatest uprising against American authority yet experienced was coming to a head in Sulu. The Moros had formed a coalition after ten months of dissatisfaction with the cedula and tax laws, and had entrenched on the summit of Bud Dajo. One thousand Mohammedans had fortified the extinct crater on the mountain top, defying all law and order. By March 2, it was apparent that only force could dislodge the Moros, and General Wood gave orders for the summit to be taken by assault. A force of 738 Scouts and regulars, and 51 men of the Sulu and Zamboanga Constabulary moved to the attack. Captain John R. White led the Constabulary, and in a desperate charge on the *cottas* was severely wounded. For this engagement he was awarded the Medal of Valor.

To many observers, the battle of Bud Dajo remains a blot on the American military occupation of Sulu. The engagement was strictly an Army attack, supported by a small detachment of Constabulary. In this largest battle of the Moro wars, 994 of the defending Moros were slain; but six Moros left the mountain at the close of the battle. The Constabulary detachment suffered the highest percentage of casualties of any troops engaged, 40 per cent of the men and officers being killed or wounded. With Captain White, Sergeant Arcadio Alga was awarded the Medal of Valor.

Although great criticism was directed against the American military authorities for this slaughter of natives, the facts seem to indicate that General Wood had no alternative. When the Joloano Moro sought a mountain top, he did so with the expectation of a fight to the finish. With their women and children about them, the Moros calmly went about the business of dying, and there was no thought in their minds of compromise or surrender. It was a messy business.

But Bud Dajo, with its slaughter, was only the beginning of the Moro resistance. Following the battle in the crater, the scene of activity shifted deeper into the bush,—detached warfare, a platoon of Constabulary against a Moro *cotta*.

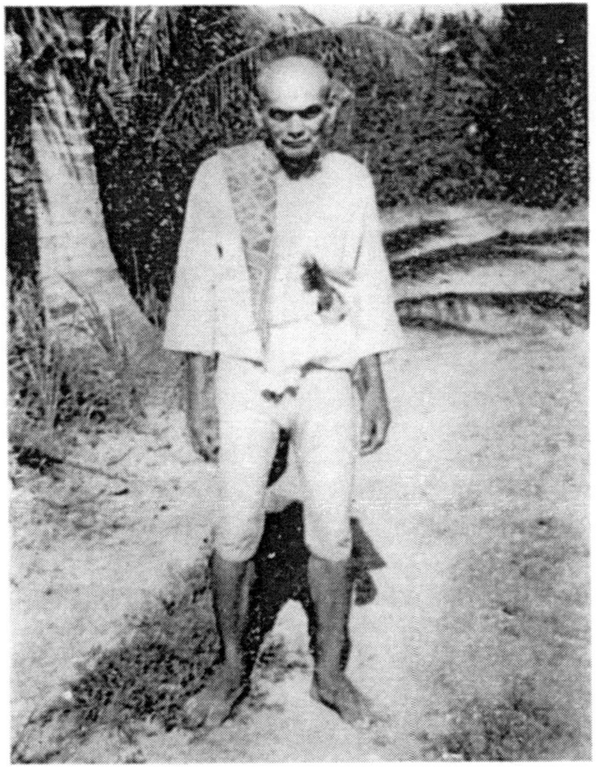

Jammang — one of the original members of Jikiri's band.

Jesse Tiffany

Moro Cotta. Note Bamboo Firing Tubes

Constabulary in possession of captured cotta

The Sulu Moro usually constructed this *cotta* on level ground, in the center of an open field of *cogon* grass. This means of defensive warfare was constructed with great military skill. The walls were of rock or packed earth, raised to a height of ten feet, and penetrated at intervals with bamboo tubes through which the Mohammedans directed rifle fire. Under the walls the Moros constructed their fire pits, protected from above by the walls. Just outside the fort and circling it, the Moros were accustomed to erect loosely woven bamboo stockades intended to delay storming troopers long enough for the defenders to pour in a withering close-range rifle fire.

The *cotta* itself was usually encircled with a deep moat, sometimes filled with water, and the only means of entrance would be long bamboo poles across the moat. As their last means of defense, the Moros had their *barongs* and krises for close-order work if the Constabulary succeeded in crossing the moat to the walls.

Over the face of the Moro country the *cottas* began to spring up by the hundred. The story of the Moro wars is one of constant assault, destruction, and rebuilding of the *cottas*.

Against the *cottas,* the Constabulary patrols went into immediate action. Among the earlier combat officers Lieutenant Leonard Furlong had no peer. He fought all over the Moro country, with stations at Davao, Cotabato, Lanao, Siasi, Cagayan, Dansalan, and elsewhere. His record shows six years of combat without a furlough, and he was admitted to have been one of the greatest combat officers to see service in the Philippines.

As early as 1904, this slight officer of Constabulary had acquired a reputation as a relentless fighter. It was said of him that he would go with half a dozen of his Moro Constabulary where a company of infantry would hesitate to penetrate. In July, 1904, Furlong had been a Second Lieutenant of the Cotabato Constabulary on station at Kudarangan. There were also three companies of infantry and two of Scouts at Kudarangan, and every time they ventured from the camp the Moros would kill one or two and make away, unhurt, with rifles. General Wood had requested that a scouting party be sent out to determine the strength of the Moros. The country around Kudarangan was all swamp, with *tigbao* grass ten to twenty feet in height, networked by narrow trails. It was a terrain where a dozen men who knew the trails could lie in wait and massacre a hundred. Only a few weeks previous to Wood's request for a scouting party, Company F of the 17th Infantry had been slashed by the Moros, losing two officers and seventeen men out of a company of thirty-six, without ever seeing the attacking force.

Leonard Furlong took over the scouting detail for General Wood. With fourteen Constabulary (all new recruits less than two months in the service), he penetrated seven miles into the heart of the territory of *Datu* Ali's best general, one Bapa ni Manakup; killed the Moros who opposed him; captured several rifles; lost two men killed and *brought their bodies back;* and fought to the very edge of the encampment at Kudarangan. His service record is filled with dozens of such back-to-the-wall victories.

Furlong was loved and feared by his Moros. He was a remarkable revolver shot, and was the best map maker in the Constabulary; he had a camera eye for details of terrain. In appearance he was slight and dark, and of seeming frailty, but his endurance on the trail was not surpassed by any officer of the corps. His recklessness in action gave him the reputation of bearing a charmed life. Today, he is known to the Moros as the possessor of a charm against bullets in battle.

His custom was to enter a *cotta* far in advance of his men, there to go berserk at close quarters. He was one of the few men of all the magnificent ones who saw service in the Philippines who was credited with supernatural powers by the Moros, and *feared* as an unearthly being.

With his fierce Moro soldiers at his back, Furlong would approach the walls of a *cotta* filled with hostile Moros. Sailing his campaign hat over the walls he went into action, with the understanding that the retriever of the hat became its owner. His soldiers vied with him in scrambling over the walls to recover the headpiece, but it is of record that Furlong never lost his hat.

He is the only man of the Philippine Constabulary to be recommended for the Medal of Valor on four different occasions. In July, 1906, he was sent into the field with orders to apprehend or kill the murderers of Private McDonald of the 19th Infantry. Furlong received word that the killers (a force of Moros commanded by the Sultan Dimbara) were at the *barrio* of Bugasan in Cotabato, Mindanao. With five Constabulary, two Scout soldiers and four Moros, Furlong arrived at Bugasan at daylight on the morning of July 9. He had but six rifles in his party. He called to the inhabitants of the house to surrender, and found, not a few Moros, but a gang of 100 armed bandits who surrounded his small force. In one of the most dramatic hand-to-hand combats of the period, Furlong personally killed six of the Moros, and extricated his men without injury to his force. He personally broke a passage through a wall of krismen as point of that compact group of soldiers who battled hand to hand with the odds ten to one against them.

One of the most striking examples of Furlong's policing activities was his extermination of *Kali* Pandopatan, the Sultan of Buldung. The *Kali* had been playing double with the

American government, and Furlong, with a dozen Constabulary, had gone to the *cotta* of the Kali for a conference. Once inside the *cotta,* he was set upon by more than 400 Moros, armed with *barongs*. Furlong backed his party into an angle of the walls and was in possession of the field after a terrible hour of slaughter.

From April 28 until July 5, 1907, Furlong was on the extended campaign that culminated in his winning the Medal of Valor. On April 28, records show him in the Taraca Valley of Lanao in an advance on a *cotta*. As the men crossed the open country before the *cotta* walls, First Lieutenant James L. Wood received a bullet through the left thigh which knocked him to earth directly in the path of advancing *barong* men. Corporal Malaco, at his side, stood over the fallen officer and beat off the attackers until the hostile fire was silenced by an advance that deployed to the right and left of the trail. Furlong was leading with his customary dash.

A few weeks later, while in the same vicinity, Furlong's men, divided into two columns, were fired upon from a *cotta* on the left flank. Furlong detached one column and led the advance through a thicket of bamboo. The *cotta* was found to be defended by a wall fifteen feet in height, protected on three sides by a deep moat and on the fourth by the banks of the Rumayas River. The wall itself was shielded by a chevaux-de-frise of sharpened bamboo stakes.

Furlong's summons to surrender was greeted by taunts from the Moros and a renewal of the rifle fire. The gateway of the fort was chosen as the point of attack, and Furlong, with Corporal Malaco and four privates, attempted to break it in with a log. They were fired directly upon by the defenders, with the result of the death of one soldier and the wounding of two others. The attacking force made the top of the walls, but were unable to gain the inside of the *cotta*.

A retirement was made under heavy fire and the fort was examined for a more favorable point of attack. A private crawled through the underbrush and located a small opening in the chevaux-de-frise on the river side.

A general assault was made, with Furlong first through the opening and into the fort. Seventeen Moros were killed, and Furlong suffered three additional casualties in his force.

By a General Order of September 2, 1907, Malaco and Furlong were awarded the Medal of Valor.

During 1906 and 1907 the names of other great fighting men began to fill the combat reports of the Moro wars. Furlong's first rank as fighting genius of the corps began to be disputed by Wood, Tarbell, Merrill, Fultin, Whitney, Tiffany, Bell, and Preuss. Under the leadership of these officers the patrols spread across Mindanao and Sulu.

The resistance of the flaming *cottas* in 1906 and 1907 set up a specialized variety of warfare. The tactics usable against *pulajan* in Samar did not apply in Mindanao and Sulu. The Constabulary were obliged to develop a special attack and defense against the *juramentado* Moro. With few exceptions, the Moros made no massed attacks . . . their strategy was stealthy night attacks of maniacal krismen, or silent penetration of the camps for weapons. Only in times of great emotional pressure did the Moros seek their mountain tops and offer up a quality of organized and sustained resistance. Their tactics were hit and run, with advantage taken of every natural condition of the terrain. They flowed through the jungle in small bands of ten to a hundred men; often they killed and pillaged a column with the despairing soldiers allowed no glimpse of their attackers.

Against the Moros the Constabulary used no advance guards, or points, to warn of ambush. Had they sent scouts ahead of the main column they would have been slashed to death and their rifles stolen long before the main body would be aware of the attack. In Moroland the men traveled in a compact group, and they walked stolidly into ambush with no warning and no preparation for the repelling of the attackers.

The Constabulary inheritance in Mindanao and Sulu was 45,000 square miles of jungle, peopled by Mohammedan tribesmen of great military ingenuity. The story of the conquest of the southern islands is one of three decades of guerrilla warfare.

Chapter XV

MAGNIFICENT YOUNGSTERS

"Patrols should often make use of some ruse such as starting in one direction and afterward doubling on their trail, in order to deceive accomplices of ladrones who may be watching them."

—*Constabulary Manual*

THE story of the conquest of the Philippines is also a chronicle that is concerned with the lives and deaths of Magnificent Youngsters. There in the Philippine jungle the old law of the bolo — that decrees that the man with the longest reach shall survive — was the final measure of any man. By the law of the bolo these youngsters grew full man-size—and yet the grisly combat records show that they were hardly more than martial infants.

The balance between survival and a grave in the Samar jungle was delicate and expressive. "Kill or be killed"— that was the code of the bush. And death was ever present — waiting to be dealt by corded krismen, by maniacal *pulajan,* or by cobra and malaria.

And so glittering youngsters, half a world from home, became the instruments, as usual, for graybeards at Washington, D. C. The young officers of that poorly paid, poorly equipped force, the Philippine Constabulary, were not barracks soldiers. They were patrolmen — and that black jungle was their beat. Their domain and their service precluded the possibility of the companionship of a wife. Indeed, the junior officers who led the patrols were forbidden to marry until they rated a Captaincy. Sometimes they came from their own jungle world to another world where U. S. regulars danced with their wives at post formals . . . it was then that the young

Constabularymen realized to the full the antisocial aspects of chasing Moros.

White women were something precious and perfect and wholly unattainable to these bush youngsters as they came in, dazed from solitude, after months in the jungle. Occasionally they attended dinner parties to enjoy, for an evening, the conversation of a woman of their own race. Beaten by the bush and monosyllabic in speech, their very incoherence was at once amusing and pathetic.

These young Constabulary lieutenants could drink all day and never become drunk at all. In like manner, they could apparently fight all day and never experience the chill emotion of fear. Their monumental drinking awes an investigator—for most of them it was the one surcease from slaughter.

Oscar Preuss, the eminent Moro killer, lived several lifetimes in the span of his short, hard-bitten life. As a fighter he was a cold-blooded genius — as a drinking man he was super-colossal. At 4:30 in the afternoon he began on a quart of Gordon's Gin — at midnight it was finished and Preuss was deadly sober. He was a great soldier — and almost too rough for Mindanao. His career had included a term as a Sergeant in the German Lancers during the Boxer Rebellion in China. He had then crossed to East Africa as a Lieutenant of Infantry. Various South and Central American revolutions saw him in action, and he had ridden for Uncle Sam as a cavalryman.

He made few military mistakes. One of them had been the time he disarmed a Moro and neglected to search the natives' hair for a dagger. He bashed out the Mohammedan's brains when the knife flashed into view, but not before the Moro had slashed the cheek of Preuss and pierced the roof of his mouth.

No man of the corps had greater stock of bush lore than

Preuss. He believed in long marches before noon — with long rests in the afternoon. The Junior officer who learned from this Prussian was fortunate. Preuss was educated, and a marvelous linguist. He spoke English without a trace of accent. His short, stocky frame seemed impervious to fatigue.

Preuss is admitted to have been the greatest Moro killer of them all. They say he was called to Manila to justify his ruthless slaughter in Mindanao. A Colonel of the Board of Inquiry questioned him, "Captain Preuss, it is said that you, personally, have killed 250 Moros. What is your statement, sir, to that report?"

Preuss drew himself up, and officers say his tone was placid and yet discontented: "The report is in error, Colonel; my count places the total at 265."

In 1911 Preuss won a Medal of Valor at Mailog Cotta in Lanao. He was then a First Lieutenant of Constabulary, with four years' service. It was his sixth or seventh war, though Preuss was then but thirty-three.

Leonard Furlong was another who piled endless years into his short span of life. He wore the Medal of Valor for the Taraca campaigns at thirty: a quick-tempered flashing fighter who was burning up inside. At fifteen Furlong was a Naval Cadet; at twenty-one he was fighting Indians in Minnesota; at twenty-two he was in the Philippines in the first American expedition to Mindanao; and at thirty-four he was dead by his own hand. He was but one of that pageant of magnificent youngsters that was the commissioned force of the Philippine Constabulary.

One of the most dashing of the officers of the days of conquest was Vernon L. Whitney. His biography, written by his mates, lies before me. "Six-footer," they said; "a big overgrown boy. Splendid rifle and pistol shot and a fine drinking man." Another mate writes, with affection in every line:

"Bad about borrowing money, spending it, and forgetting it. Liberal with his own and everybody else's money. That was Whitney and it was about his only fault."

Few men could fill Whitney's shoes figuratively—no one in the corps could hope to occupy them literally. In one of the Moro campaigns Lieutenant Tiffany had crossed over the Kulingtang mountains with a strong party of Lanao Constabulary in search of outlaws. They came to an old campground, littered with rusty cans, in an isolated section. There was great speculation as to who had made the camp. Some expressed the thought that Fort had passed that way; others considered it a former camp of Furlong or Preuss. A Constabulary private settled the argument; he emerged from the bush with a howl of laughter crying "*Teniente* Whitney." He was holding by the laces a moldy outworn shoe—size 14.

Whitney, the ham-fisted, was strong as Sandow and huge in frame. He was accustomed to swell his chest and pop all the buttons from his uniform coat. For amusement, he would put his fist through a door. He was a terror to the Moros, and after that sanguinary battle at Sahipas *cotta* in which Whitney received a bullet that severed a sciatic nerve, he was selected by General Pershing as a fitting Governor of Sulu. His Medal of Valor had been earned at twenty-six at Mailog; in 1916, when he retired, he was thirty-one years of age. Behind him were eleven years of battle!

James L. Cochrun was one of the most popular officers ever to wear the khaki and red uniform. "Gentlemen Jim" to all his mates; tall and spare, susceptible to women, a man who could pack his liquor. At twenty-one, fresh from school, he crossed the Pacific to join the Constabulary. At twenty-eight he was winning a Medal of Valor before Sahipas Cotta on the island of Jolo. Left arm smashed by a bullet, he pushed on until he fell, drilled through the abdomen. With seven years of jungle warfare behind him, he retired in 1914 on two-thirds disability. A veteran at the age of twenty-nine.

Samuel T. Polk of the fiery temper was a lank Mississippian who would fight anything—anytime. Medal of Valor at twenty-five; retired veteran at thirty.

And then there was Donald Root, youngest of all winners of the Medal of Valor. At twenty-one he was at the head of a patrol in the dense bush near Mamaya Peak, Lanao. Wounded in the left side, swarmed over by the hostile Moro force of Raja Muda Randi, this boy grown suddenly to full man-size, extricated his patrol and accomplished his mission. He resigned at twenty-six, to wander away to the war in France.

Another magnificent youngster was John R. White, who had been in the Greek Foreign Legion at eighteen, in the American infantry at twenty, in the Philippine Constabulary at twenty-two. For fourteen years he fought Moros and jungle, and his Medal of Valor was won at Bud Dajo when he was hardly more than a boy. At thirty-five, he was retired a Colonel, with a lifetime of crowded hours to mark his combat service.

The list is so long. Cary Crockett at Bulao, in Samar, winning his medal at twenty-six; Fort, a regular Daniel Boone who spoke every native dialect of Mindanao. "Old Susan," his Winchester, was notched with a record of his victims.

John Fawcett, secretive, cool, efficient and loyal, one of the best all-around officers in the Corps; Ernest Johnson, dead on a Moro spear in his twenties, a boy from the University of Oregon who failed to come home; J. C. Tiffany, an Oregon State College youngster and one of the finest fighting men in the Mohammedan country, for whom there were eight years of combat against the Moros until his retirement in 1917 to engage in the greater war in Europe.

Among the romantic youngsters who wrote their names

permanently into the records of the Philippine wars was the "Red Lieutenant," James L. Wood. Wood had been ranking cadet in his class at Culver Military Academy, and as such rated a commission as Second Lieutenant in the regular army. But life in a military post was not to his liking, and this restless fighting man turned across a far horizon to take on with the muddy riflemen of the jungle patrol.

In 1904 he came to Manila, where he became a Third Lieutenant in the Philippine Constabulary with a service assignment in Mindanao. He was remarkably equipped for his job. He spoke Spanish fluently, and during his Philippine service he learned eighteen of the native dialects. By birth, education, and ability, Wood was certainly one of the highest type men to seek action in the Constabulary.

In appearance he was striking. Six feet one inch in height, he weighed 180 pounds. Wood had been an all-around athlete, and some of his records at Culver still remain to be broken. He is credited with knocking out a former world boxing champion in a single round. He was a born horseman, sitting his mount like a Cossack, and capable of performing startling feats of trick riding.

Among the Moros, Wood was known as the "Red Lieutenant" because of his thatch of red hair, and they said of him that his hair gave off sparks when he was angry.

Wood was a ballistics expert. He had carried with him to the Islands a Winchester .45-70 rifle, complete with bullet molds and reloading equipment. As a result of this interest in preparing his own loads, he developed a unique bit of strategy that was most effective in recovering outstanding rifles in Moro hands. Major Wise of the Scouts had used the device casually in Samar, but Wood put it to positive use as an effective means of discouraging the use of rifles by the Moros.

He was ordered shortly after his arrival in Mindanao into

that endless and tiresome duty of rounding up firearms in outlaws' possession. Wood was quick to note the fact that the Moros carefully followed Constabulary details in the hope of finding lost cartridges. Even with the greatest care, bodies of men on the march were constantly losing bits of equipment. Occasionally a cartridge would fall from a loop, to be swooped upon by the Moros, for cartridges were worth their weight in gold in Mindanao and Sulu.

It was soon noted by the Moros that Wood's detail was downright careless in this matter of losing shells. A visit to his abandoned camp sites always resulted in the finding of six or eight cartridges.

And then one day, a malignantly unfriendly Datu sent an emissary to the Constabulary post stating that he desired peace and was willing to turn in a rifle. . . *in bad condition.* It seemed that the *Datu* also wished to come in for hospitalization . . . having met with a serious accident. He was brought to the post with the entire right side of his face missing. His rifle barrel was split and the entire bolt action was missing.

Rifles lost face in the vicinity of Iligan, with the result that a Krag, in first-class condition, once a priceless commodity, would no longer make even a first payment on a third-class Moro wife. Wood had seen to that. He had loaded several hundred government cartridges with dynamite, and conveniently dropped them to be found by the Moros.

Wood's first experiment in the field of ballistics encouraged him to try another. He evolved a special "Wood's bomb" for use against the *cottas*. It consisted of two tin cans, one within the other. The inner can was packed with white nitro, and the outer with carriage bolts, nuts, scrap iron, and nails. One night he had his first opportunity to try the weapon. Approaching with his men, he hurled the bomb over a *cotta* wall and waited developments. He had not long to wait . . . within ten seconds the bomb was lying sizzling at his feet, hurled back by the defenders. Hastily Wood threw it over the walls again—just in time. Thereafter he experimented in detail with fuse lengths, and in time developed a bomb that was the equal in every respect of the grenades of the doughty Conway of the Scouts.

The Moros, ever ready to learn from an ingenious foe, developed Mohammedan improvements to combat Wood. On one occasion the Lieutenant approached a *cotta* wall with a patrol, when suddenly a number of Moros appeared with a bamboo tube twelve feet in length and four inches in diameter. This had been stoutly wrapped with wire and half filled with scrap iron and powder. It was fine in theory, and the Moros lighted the slow fuse and braced the cannon with their shoulders. The Constabulary took temporary cover, and when the explosion came at last there was a shattered *cotta,* a few rifles—but no Moros.

Wood was the trail companion of Leonard Furlong in many of the latter's most stirring campaigns. On one of the campaigns in Lanao the two officers, with their respective detachments, were driving through the jungle through two converging trails, hoping to flank the Moros at a point where a bridge crossed a waterway. The Moros reached the bridge a short stride ahead of Furlong—he was able to capture one straggler.

When Wood came up with his detachment he saw a curious sight. Furlong was directing the operation of lowering the captive Moro by means of a rope attached about his middle, into the deep water beneath the bridge. The Lieutenant had seen the fleeing Moros throwing their rifles into the water and was determined to recover them, As Wood approached, Furlong looked over his shoulder, and then spoke to his Sergeant as the dripping Moro was hauled up with a rifle he had retrieved from the deep water: "All right, Sergeant, bait him up again and send him down after another one."

A great deal has been indicated of the amount of gunfire it takes to kill a Moro . . . not so much has been written about the amount of hanging a good Moro can stand. While Wood was acting as Chief Justice of the Tribal Ward Courts of Lanao Province he had occasion to sentence a Moro to be hanged. Without delay the Moro *was* hanged. . . hanged for a good long time. The attending physician pronounced the Moro dead, the clerk of the Court recorded the undeniable fact, and a member of the family of the late lamented was allowed to cut the victim down. After a few minutes of lying flat on his back to give dignity and a certain suspense to the occasion, the Moro sat up. He was officially dead, and he couldn't be hanged again unless he committed another murder or two, for legally he no longer existed. And possibly to this day he lives there in Mindanao, telling his grandchildren about that day when his breathing had been slightly impeded by process of law.

James L. Wood, the "Red Lieutenant," remains one of the fascinating characters of the age of the American conquest of the Philippines. He lived many lifetimes in his comparatively short span of years, for the Philippine wars were but a warming-up process for a strenuous career at arms.

Fabulous youngsters, living each day for the day—and careless of tomorrow. Hurrying down the beds of foaming rivers deep in the jungle; facing a bush that was flecked with blood and flame and flashing cold steel; leading their patrols against impossible odds—to return to fight again. Restless, gallant, always on the move—always in a hurry. Too often hurrying away to die.

Sometimes they gathered in their barracks and in the villages for a brief moment of relaxation with battle song before duty recalled them to the bush. Living in every sense the words of that poem by Bartholomew Dawling which is known to every officer of Constabulary. "Stand by your glasses steady."

These youngsters invested their years and the best of their lives in the Philippine jungle. Their reward has been pitifully small, for few came home unscathed. But they served their god of battle well, and possibly with that they are content.

In the Moro country five names stand out to head the list of truly great combatants. Fort, Furlong, Tiffany, Whitney, and Preuss. "Quintuplets of Death" they might have been called, for individually they killed more Moro outlaws than any other Constabulary officer in Mindanao and Sulu.

These youngsters engaged in the fiercest hand-to-hand combat that is to be found in the combat records of America. No journal of the Indian wars contains the note of horror that was injected by the *pulajans* and the Moros of the Philippines. No terrain of battle is more terrible than the Philippine jungle. It was fighting that took toll of life and limb and health—it was a job for magnificent youngsters.

Chapter xvi

THE LENGTHENED PATROL

"Night patrols will be successful only when great care is exercised in the matter of silence; lighting of matches, smoking, making noises with equipment, etc., should be avoided."

—*Constabulary Manual*

WITH the passage of the weary years of patrol, the Constabulary was growing in strength and dignity. The year 1908 saw the corps of 4,573 men directed by 315 fine jungle-bred officers. The year also saw, for the first time, complete control of the Constabulary in Insular affairs. Not in a single instance did the regulars or the Scouts take hand in the preservation of the peace of the Philippines. Patrols traversed more than 230,000 miles of jungle trails, but the preservation of the peace was a weary task.

The northern islands were quiescent, but not quiet. Forty-four heads were taken in the mountain province of Lepanto-Bontoc . . . civilization came slowly to the head-hunters of northern Luzon.

The Insular government was again bubbling with optimism. The report for the year 1908 says: "It is eminently gratifying to state that from July 1, 1908, to date (November 3, 1908), a state of complete peace has existed throughout the Archipelago—*with the exception of some slight disorders caused by raids of lawless Datus and brigands in the Moro Province.*"

The note of optimism was not sustained by the happenings of the year. Harry M. Ickes of the Bureau of Science was killed in Mindanao; a few weeks later, H. E. Everett and Tilden R. Wakely, with a party of three, were ambushed in

Negros; Felipe Salvador was still at large on Samar; and Jikiri, greatest of all the Moro leaders, was cutting a swath of blood and terror in Moroland. In Lanao Province, in Mindanao, the "lawless brigands" of the report were composed of five well-armed Moro bands, fortified in well-defended *cottas*.

There was no question but that the rebellious factions were sputtering into extinction, except in Mindanao. Capable Constabulary officers in the north had seen to that. On Cebu was Matteo Luga, who had been a gallant Insurgent leader before donning the red epaulets of the Constabulary. No member of the corps wore the uniform of the jungle police with greater distinction than this swart, fearless Filipino.

Luga had fought Stacey in the middle islands, and there were other American officers to testify to the courage and cunning of this leader. He was an honorable soldier who earned the respect of the American army. They tell a tale about Luga. It happened during the days of the insurrection, when he had been one of Aguinaldo's most sturdy commanders. Filipino soldiers serving the American cause had been inflamed with the legends of the mighty Luga, and had deserted the American camp and sought service under their countrymen. Luga had heard them out and then had placed them under an armed guard. At daylight a small squad of men had appeared before the American lines. With them were the deserters and a message to the American commander from Luga. He had written, "I return to you deserters from your camp who sought service with me. I request that you do the same should any of my men weaken from their duty. It is my desire to wage honorable warfare."

That then, was Luga, who in 1908 was wearing the red epaulets of the Constabulary. He had been sent to Cebu Island with instructions to keep the island clean and bandit free. And Cebu *was* clean, under the administration of this capable and valiant native officer.

Before me lies Luga's accolade, written by an American officer who saw much service in the Philippines: "In Matteo Luga, you saw a man to remember as long as you lived."

As public confidence in the Constabulary had increased, the corps benefitted by arms and equipment. The old fear of arming native soldiers was fading now, and 3,365 Krag rifles came into possession of the force. With but eighteen desertions in the year 1908, the Constabulary was in high morale as it turned to the final battleground in the Philippines.

The army was beginning to give way in Mindanao and Sulu, as the Constabulary grew into power and needed strength. The Krag of the Constabulary moved in to replace the Springfield of the regulars. With the coming of the Constabulary the mode of warfare began to change. The massed troop movements gave way to the familiar system of patrols; the rapid fire was displaced by the rifles of steady individual marksmen.

The Moros deep in the hills and swamps were beyond the reach of any law save by that old method of infiltration by means of the patrol. And that was a manner of warfare that was foreign to the practice or the inclination of the American regular.

In Mindanao, old Ampuan Agaos and Amai-gin-dalungan were commanding a force of more than 400 men, armed with 134 rifles. It was against these leaders that the patrols were directed. In January, Lieutenant Wood and a detachment went against Amai-gin-dalungan on the Romain River, forcing the Moros deeper into the hills. Two weeks later Amai came back with a force of eighty Moros armed with rifles, and made a bloody attack on the Constabulary headquarters at Dansalan.

The records show the Constabulary in daily action, with Wood, Tarbell, Hemmett, Fulton, Whitney, and Fort continually in the field. In April, Wood attacked Ampuanagaos in a *cotta* near Iligan, supported by men of the 18th Infantry under Lieutenant Endsley. Fort fought a red day at the same place a few weeks later, killing thirty-three Moros of the force of Amai. Whitney and Fulton repeated the hammer blows near Tugaya, lessening the outlaw strength by twenty-four. In each encounter the Constabulary casualties were severe. The Moros were warming to the encounter, testing the strength of the Insular Police.

And then Captain Elarth took ten men against a thousand! He was in the hills of Mindanao, investigating a report of Moro organization, and he came into contact with a thousand tribesmen, armed and ready for action. As a Constabulary officer, he was supposed to do something—and he did. He called for a parley with the headmen; and the Constabulary—ten men and the Captain—sat down on the summit of a hill, surrounded by the hillmen. Three Moros on the edge of the crowd began to mutter and the headmen rose from the ground and began to draw away. Then the trio of frenzied fanatics drew their weapons and rushed the Constabulary Captain. The Constabulary took refuge in a rally formation, with fixed bayonets. The leading Moro was almost upon them before Elarth could draw his pistol. "*Pot-i-na*" (Die now): the voice of the Moro was a scream as he hurled himself upon the Captain. At the same instant the hillmen released a shower of spears.

Elarth dropped the first two Moros with skull shots from his pistol, but there was no time to stop the third, who was armed with a spear. There was a movement behind the doomed Captain, and Sergeant Alvarez leaped forward to take the spear in his chest. Too late to save his Sergeant, Elarth blew the Moro's head away with a .45 calibre bullet.

Had the long-haired hillmen supported the three Moro leaders as they charged, the entire detachment would have been wiped out with the loss of eleven rifles. But the hillmen contented themselves with showers of spears before they melted into the jungle. Left on the field were eight dead Constabulary bodies bristling with spears.

Elarth, with his two surviving men, jerked the bolts from the dead men's rifles and plunged into the deep bush. All day and all night they marched, to return safely to the Constabulary post. Elarth had ably upheld that old fighting tradition of the Corps: "To be outnumbered always; to be outfought, never."

In 1908 occurred the last case of desertion in the Lanao companies. The 2nd Lanao had long been on station at Zamboanga on guard duty, and the morale of the men was low. At an outpost on Lake Dapao was one Sergeant Romandiar who had grown tired of the discipline of military life and longed for a career of banditry One dark night Romandiar, a Corporal, and four privates passed through the sentry lines and took to the hills. They carried with them their rifles and a plentiful supply of ammunition.

Overnight, the band became full-fledged outlaws. In the weeks that followed they roamed all of the Dapao and Lake Nunungan country, eluding the Constabulary patrols who sought to apprehend them.

It was then that Captain Preuss; their old commander, returned to duty from an extended vacation. After chasing the men through the Nunungan mountains for some time with no result, Preuss suddenly threw a line of skirmishers across a district that was heavily populated with Moros under influential *Datus*. The Constabulary laid a path of rifle fire before them as they began a slow advance. The *Datus* came under a white flag to beg that Preuss cease firing into the villages. Preuss replied, "I am here for the heads of six deserters from my Lanao Company. By helping them escape, you have declared yourselves outlaws. Not a man in my path will be spared until the heads of those deserters are brought to me in a sack."

And brought they were, within a period of a few hours, to end desertion in the Lanao Constabulary.

On June 6, 1909, occurred the famous Davao Mutiny, which was to reflect discredit upon the jungle police. Davao, in the center of the hemp region in southeastern Mindanao, was garrisoned lightly with Constabulary, mostly engaged in the settlement of intertribal disputes among the hillmen. The constant, bloody action that was the lot of the Lanao and Cotobato posts was lacking in this quiet station, and the morale of the men suffered through lack of activity.

At ten o'clock of the evening of June 6, the forty-eight men in the post were quiet in barracks, when, at a prearranged signal, twenty-three Constabulary privates seized arms and broke from the barracks. Krag bullets spattered against the walls of the buildings of Davao as the factions fought through the streets. The firing was heavy, but the casualties were small. One American civilian was killed, trapped in the line of fire, and four loyal Constabularymen were wounded.

Lieutenant H. H. Noble rushed to the scene with a detachment of the 23rd Infantry, and Davao was placed under military law. All of the mutineers save five were captured in a brisk round-up that required the attention of the regulars for several days. No real damage resulted from the Davao mutiny save the loss of a few rifles, but the break came as a most unwelcome surprise to the Administration.

The old doubts were revived, and the army made capital of the fact that they had been called in to restore order. But public opinion could not be crystallized against this Insular Police force that had so successfully waged combat for a terrible decade. The re-arming measures proceeded, and the Constabulary built up to the years of Moro fighting that remained to their lot.

The casualty figures of regular army and Scouts for the years 1909 and 1910 are interesting. During this fierce era of *cotta* fighting against the Moros in Mindanao and Sulu, the army records show four regular soldiers and one Scout killed in action in 1909, and four regulars and no Scouts killed in action in 1910. Possibly indication enough of the relative jungle service of the Police and the Army.

We turn back the months now, to the year 1907, to witness the beginning of the career of the most terrible bandit that the Moros were to produce. In July of that year a young betel-nut bearer of the Sultan of Sulu gathered about him seven pirates of ability and set out on a patrol of terror. Jikiri was his name. Today he is a legend of the Moros, and with reason.

There was nothing in the career of Jikiri to warrant his elevation to the rank of hero unless it was that magnificent stand he made against the combined forces of army, Constabulary, and Scouts that reached across a period of more than two years. He was a wraith who struck and retreated, and he handled his campaign of death and looting and rapine with the skill of a professional soldier. With the possible exception of the *Datu* Ali in Cotobato, Jikiri was the ablest war leader produced by the Moros. He had all the qualifications for leadership. A great personal magnetism, the cunning of a leopard, the ferocity of a boar, and the benefit of Arab blood to give him prestige. He was tall and broad-shouldered for a Malay, with a hooked nose and a flair for fancy dress. He went into action with a white cloth draped across his face, and he was feared by American and Malay from the coast of Borneo to the north of Mindanao.

The night of November 1, 1907, a Chinese trader named Tao Tila had the dubious distinction of being the first recorded victim of Jikiri. The Chinese was sailing a *vinta* along the coast of the island of Jolo, engaged in trade with the Moros. Off the coast of Lumapid, in the blackness of night, a swift sailing boat sped out of the dark, and a voice aboard the Malay privateer called in the Sulu tongue, "Kill them."

A moment later the pirate ship was alongside, and the crew of the Chinese boat were stricken with krises before they could rise from their benches.

With the proceeds of this piratical raid Jikiri clothed his men in distinctive garb, and a few weeks later, on Christmas Eve, he raided the American lumber camp at Kopuga, on Basilan Island.

In the raid on Kopuga, Jikiri demonstrated a cold-blooded ferocity that showed the Malay at his worst in the Philippines.

As we turn back to that day we see the two boats of Jikiri arriving at a point near the camp at two o'clock in the afternoon. The camp laborers have been paid off, and the camp is deserted except for Case and Verment, the loggers, Mrs. Case, her mother-in-law, a Moro woman, and a native foreman. Jikiri sent two men to reconnoiter. They entered the camp and approached Case, offering to purchase a *vinta*. Case replied that they had no boats to sell, and the Moros withdrew.

At five o'clock the raid began. The seven Moros deployed

about the camp. On signal, one of the bandits entered the store where Mrs. Case was arranging the stock and asked for cigarettes. As the woman turned to the shelves she heard Verment scream outside and, looking through the window, saw the logger go down before the blades of two Moros.

The Moro in the store leaped for the woman, but the high counter was in his path and she managed to escape through the window and make her way to a Yakan village.

As Verment lay dying outside the store, Case was set upon by two other Moros, who severed his head with a stroke. The wife of the dead Verment received a ghastly kris wound that laid open her back from shoulder to hip.

After twelve dreadful hours in a *vinta* manned by Yakan sailors, the survivors reached Zamboanga, where Mrs. Verment died from her terrible wounds.

General Tasker Bliss, on station at Zamboanga, sent out an immediate patrol under Lieutenant Shutan, who found the bodies of Case and Verment, and scoured the countryside with no success.

A week later, reports began to drift into Constabulary headquarters from the island of Sibago, where Jikiri and his men were spending freely and talking of the raid on Kopuga. When patrols arrived he was gone and his gang were disbanded to wait developments. Under orders from Jikiri a rendezvous was made on Patian Island, on the fringe of Jolo, and there the eight desperados met in January, 1908, to lay plans for the future. Captain Newbold went in pursuit with 200 men, but Jikiri went into hiding in the swamps of Lumapid and for three months he played tag with the finest American forces in the southern islands.

In April, 1908, a troop of the 6th U. S. Cavalry were riding the rough country near Jolo as part of that ceaseless patrol after Jikri. In a ravine heavily wooded, they found him in position on a commanding slope. There was nothing for the cavalrymen to fight, save the white smoke from the Remingtons of the outlaws, and Jikiri was away again after dropping Trooper Ferguson with a well-placed shot.

The swamps closed behind the outlaw; and then the swamps opened again and he came out to waylay Albert Burleigh, an American schoolteacher stationed at Maybun, on Jolo Island. Burleigh was ambushed on a lonely trail and cut into minute pieces by the krismen of Jikiri.

It was time for concerted, vigorous action against Jikiri, who was the most serious menace in Moroland to American prestige. The *Mindanao Herald* opened an editorial campaign, and military conferences were held to discuss the elimination of Jikiri. The *Herald,* in August, 1908, published this statement: "Jikiri has evaded the authorities for so long that the Moros are beginning to entertain a great respect for his prowess, and unless he is accounted for soon, he will be the cause of serious disorder."

The Sultan of Sulu entered into direct co-operation with American forces. The personal kris guard of the Sultan took the field, and Hadji Butu, Prime Minister of the Sultan, organized an official anti-Jikiri force which went into immediate patrol. But Jikiri was not to be taken so easily. His force had grown to a formidable, well-armed gang, and he was credited with at least forty successful raids. He made plans now for an ambitious career of piracy.

The sinking of a pearling schooner was his next objective. The night of August 22, 1908, this pearler was at anchor off Tonquil Island, making preparations for the return voyage to Zamboanga. The hold contained more than a half-ton of black-lipped shell. Six of the crew were asleep beneath deck, and it was near midnight when the lookout heard the approach of sailing *vintas*. Before he could hail the boats, a rifle shot from the night dropped him dead on the deck. As the six off-duty sailors rushed to the deck, the pirate boat came alongside the pearler and kris blades flashed in the moonlight. In a terrible, bloody half-minute four men died, and the cook and one badly slashed sailor escaped over the side to swim to safety.

The pirate ship then made off with the loot. The next morning fishermen found the silent pearler with five dead men on the decks.

Jikiri's force had now grown to 100 men and he entered piracy on a scale that had not been seen in Sulu since the days of the Moro power in the sixteenth century. In January, 1909, the last year of his life, Jikiri opened operations with a massed attack on four pearlers. As the slim outrigger pirate boats bore down on the pearling fleet, two of the luggers made sail and escaped to Zamboanga. The remaining two, the *Ida* and the *Nancy,* were attacked at long range by rifle fire and portable artillery. Four pirate boats circled the pearlers, and a sea battle that lasted for three hours was waged before the *Ida* ran out of ammunition. Their guns empty, the crew of the *Ida* leaped overboard and swam to the safety of the bush. From there, they watched the looting of the vessel. The *Nancy,* meanwhile, was under attack from three pirate vessels, and the marksmen on the pearler were able to inflict severe casualties as the pirates drove to closer quarters. Jikiri himself gave the orders for that last frenzy of preparatory gun-fire, watching with veiled face as the bullets swept the deck of the *Nancy* and thudded into the planking below waterline. As the crew took refuge behind the bulwarks, the pirate boats surged in to come alongside, and the Moros leaped to hand-to-hand combat on the deck of the pearler. The four survivors on the *Nancy* went down in that savage shouting rush.

THE LENGTHENED PATROL

The next day the government cutter *Atlanta* arrived on the scene with Captain De Witt and Lieutenant Byram and a Constabulary detachment. The pirates were still alongside, removing the last bits of equipment from the dismantled and burning *Nancy*. Seeing the approach of the Constabulary cutter, Jikiri ordered a withdrawal, and the pirates took to their speedy craft and escaped without difficulty. The *Nancy* was fired by the Constabulary, and towed to sea to be sunk off Lagason.

This same Captain De Witt had permanent orders to keep his detail in pursuit of Jikiri. De Witt had been selected for that responsibility as being the best equipped officer of the district for the job as he spoke the Joloano dialect fluently. His tactics were to operate continuously with a small mobile force in the hope of catching Jikiri off guard. He came upon Jikiri near Parang, in Jolo, shortly after the sinking of the *Nancy*, and after a savage fight resulting in the deaths of four of the pirates, Jikiri again made his escape into the swamps near Lumapid.

He was then reported in British North Borneo, and two governments collaborated in search for him. Field forces of the North Borneo Constabulary went after him and succeeded in killing several of his band. Jikiri himself escaped in a *vinta* and returned across the strait to his old hangouts near Jolo.

Within a month of the sinking of the *Nancy*, Jikiri had claimed additional white victims. The Constabulary post at Siasi was assaulted in search of additional arms and ammunition. More than 600 shots were fired before the pirates were beaten off. Jikiri stood in a boat before the post, directing operations, with a white cloth concealing his features. In this engagement, M. H. Holmes, an American planter, was among the killed, and the pirates left four dead on the beach.

Three days later, Jikiri was eighty-five miles down the coast, attacking the town of Tugig-Indangan, which was the landing place of the first Mohammedan missionary in the Philippines in the year 1380. Lieutenant Hasemeyer arrived too late to prevent the sack of Tubig-Indangan. He found on the beach the bodies of two more white men who had fallen before the kris blades. Wolf, an English planter, was found slashed into thirty-two pieces; his partner, Cornell, was dreadfully mutilated.

A very doughty Constabulary officer now made his bid for Jikiri. Lieutenant Peake, in company with Hadji Usman's force of Moro guardsmen, boarded a suspicious *vinta* on the Tawi-Tawi coast. Believing the crew to be members of Jikiri's band, it was decided to remove them to Bongao for questioning. All went well until the boat was in the shallow water, approaching the anchorage at Bongao. Then, with a scream, the Moro pirates made a break for liberty. In the mêlée, the boat overturned and one of the most savage battles of the campaign was fought hand to hand, in the shallow water. Peake received a pistol bullet in the leg, and went down with pirates clutching his throat in an attempt to drown him. Hadji Usman shot Peake free from his attackers, dropping two with well-directed pistol shots and beating off the others. Peake, bloody and half drowned, stood up in the water then, and with pirates all about him, gave an astounding exhibition of coolness in action. As deliberately as if on a target range he drew bead on his attackers, dropping seven men with seven consecutive shots.

In May, 1909, Jikiri withdrew the remainder of his force into the swamps of Lumapid again and sat down to plan his last campaign. His force had been riddled by the constant pounding. With thirty men about him, he launched out again on a mad trail of destruction and death. Preceding him, he sent his messages to the American government, warning that "the fighting would not cease until no man of my men can longer bear a kris." He set as his goal the personal slaughter of a hundred men, and no man today knows how nearly Jikiri made good that threat.

He bobbed up at Lake Seit against Captain Rhodes, losing four men in a bitter afternoon struggle. Captain Byram cornered him on Patian Island, but Jikiri escaped. A few days later Byram, in collaboration with Captain Signor's gunboats, fought a savage battle with the outlaw on land and sea. Five more pirates were killed.

On June 30, 1909, Jikiri was at Maybun, and Captain Byram hurried thither with two squadrons of cavalry. As the horses thundered into the little town they saw the pirate *vintas* putting to sea in the direction of Patian Island.

Jikiri, at this point, seems to have been satisfied with a career of death eclipsing that of any of his fellows. He had reached for and received the ultimate in favorable recognition as a super-bandit. And with that, Jikiri, being a Moro, was content. In the opinion of the old-timers in the Philippines who witnessed Jikiri's spectacular career, he could have escaped the American forces as easily this day in June as he had in the preceding two years.

But Jikiri was willing and ready to die.

He held a last conference with his men . . . eight hardy krismen who were ready too to ascend the weary steps to Paradise. On Patian Island, the scene of some of his greatest triumphs, he took refuge in a cave within a volcanic crater. There he waited patiently for the approach of the government force.

Troops A, B, and D of the 6th Cavalry, supported by

quick-firing guns from Signor's gunboats, closed in for the kill. In the cave with Jikiri were several women. They were given the opportunity to leave, unharmed, before the assault began. All but two of the women took advantage of the offer. Jikiri's wife and one other Moro woman elected to die with the pirates.

For two days and two nights the riflemen and quick firers poured volleys into the mouth of that cave. The smoke eddied and billowed. Jikiri was going out in style . . . with three troops of cavalry and a battery of field guns opposing nine men armed with edged blades.

It must have flattered the fighting ego of that remarkable, bloodthirsty bandit as he sat there in the crumbling cave under that terrific fire. It was the ultimate in cautious warfare that American troops waged that bloody week-end. There was to be no slip . . . Jikiri was to be killed . . . with no American casualties.

But the moment that Jikiri had lived for all his life was soon to come. Regardless of his misdemeanors against the law and his record of cold-blooded murder, there was something terribly magnificent about that Moro outlaw who terrorized the combined forces engaged in the pacification of Mindanao and Sulu.

There in his cave . . . he waited.

On the morning of July 4, 1909, it was decided, after lengthy conference, that nothing could remain alive in the cave in the face of that terrible fire. As the smoke and dust eddied from the cave mouth, American soldiers fixed bayonets and began a slow advance on the outlaw position. No shots came from the cave mouth, and they were reassured. The mountain guns were silent as the troops moved in . . . closer and closer.

And then Jikiri came.

From that smoke and dust-choked cave came eight Moros, Jikiri at their head. Scorning the odds, the hasty rifle fire, and the bayonet points, they hurled themselves upon the startled attackers. The Krags began to roar at point-blank range. Spots of blood appeared on the tunics of the Moros but they stayed on their feet. The terrible krises were upraised . . . .

Jikiri caught Lieutenant Wilson by the hair and raised his kris for decapitation. Lieutenant Bear rushed in and blew Jikiri's head into a bloody pulp with a full charge of buckshot. The kris blade deflected as Jikiri died, and Wilson fell with the pirate, blood streaming from a terrible wound.

Behind the fallen leader his eight Moros were hand to hand with the cavalrymen. Eyes were gouged out . . . heads fell from shoulders to roll away in the bright sunlight . . . men were severed in twain as the eight Moros sought the rewards of Paradise. And then suddenly, the little valley was still and the shattered American force took stock. Lieutenants Miller, Wilson, and Kennedy were wounded. Twenty soldiers were stretched on the ground with kris slashes . . . a sailor from the gun crew was slashed from neck to waist. Beneath the body of Jikiri, Private McConnell was slashed out of resemblance to human form.

Jikiri was dead . . . but he was well equipped for the weary ascent to Paradise.

The Moros are incredible. No word picture could paint, in true colors, the ferocity and inherent fighting ability of these Mohammedans of the southern Philippines. As fighting men, they have no superior in any breed of men in the world, and it is as fighting men that we should judge them.

Their terrific valor enriches and strengthens the reputations of the fighting Insular Police who opposed them.

Chapter xvii

DEATH ON THE KRIS

"Punitive expeditions against non-Christian tribes on the part of the Constabulary are forbidden without specific orders from the District Chief. It is not the policy of the government to punish a whole people or village for the deeds of individual miscreants; but with the exercise of tact and skill, to secure . . ."

—*Constabulary Manual*

THE battle phase was now definitely Constabulary against Moros. The finest fighting brains and skill of the Insular Police were directed to Mindanao and Sulu in these closing days of the first decade of service. In 1910, in the Department of Mindanao, were forty-two officers of Constabulary, spread in a thin line across the centers of disturbance.

In Manila, the administration had decided to aid the Insular Police patrols with supporting forces of units of the United States Navy. It had been learned by bitter experience in the field that the Moro was an amphibian. If he was worsted in a jungle fight he would take to sea, out of range of Constabulary activities.

To combat this situation, a "Mosquito Fleet" was manned in Manila under the designation of the "Southern Philippine Patrol." Four gunboats, the *Quiros, Pampanga, Paragua,* and *Mindoro* were commissioned, with Lieutenant Schoenfelt as Squadron Commander. With the *Quiros* as flagship, the squadron set sail for Moroland, where they were to render valuable service.

The little gunboats were deadly destructive weapons against the pirate boats. They were shallow-draft vessels, and could pursue the freebooters across the coral reefs. And they were formidably armed and manned. The armament consisted of two three-pounders—one forward on the forecastle and one on the fantail. In addition, they mounted two one-pounders in sponsons amidships. On the bridge were two Colt automatic rapid-fire guns, using Krag ammunition.

The landing parties were equipped with .38 revolvers, Krag rifles, shotguns, and cutlasses. Each ship carried a crew of thirty-five, with fifteen Filipinos in addition.

This naval detachment operated in close conjunction with the Constabulary patrols. With the location of a party of armed Moros, the Constabulary would dispatch a patrol into the bush and drive out the Moros, who would take to their boats to be run down and sunk by the Naval Patrol. This policy was successfully carried out in September, 1911, on Sitanki Island, at the extreme southern tip of the Philippine Archipelago. Sitanki had been for centuries a pirate rendezvous and a graveyard of ships. Navy and Constabulary, working in co-operation, eliminated the pirate nests in a series of brilliant operations.

During that same month, the *Pampanga* was lying at anchor at Tablas on Basilan Island, waiting the results of a shore party sent against the Moros. Ensign Hovey was leading the detachment. High in the hills near Tablas, Hovey led his men into a Moro village after an exhausting day on the trail. The outlaws he sought had melted into the bush, and the Ensign had stopped at a *nipa* hut to ask for a drink of water. As a Moro held out a cup to him, the *nipa* shack exploded with a blast of flame and Hovey took the full charge of a brass cannon. His men fought their way out as the village caught fire from the force of the exploding cannon.

Fifteen miles across the straits, at Zamboanga, the *Quiros,* sister ship of the *Pampanga,* saw the very lights streaking to the sky, and hastened to Tablas. A naval detachment entered the wild country and scoured the jungle for the vanished outlaws, returning with no result other than the recovery of the body of Ensign Hovey.

It was men from the *Quiros,* too, who witnessed the death of Lieutenant Rodney at the hands of a Moro *juramentado* in Jolo on April 16, 1911. Lieutenant Rodney, an officer of the 2nd Cavalry, had gone for a Sunday afternoon walk with his small daughter. Walking unarmed on the Jolo-Asturias road, Rodney had been preceding a seaman named Steel and two other sailors from the *Quiros* by a few steps. Before the sailors could draw their weapons, a Moro burst suddenly into view, hacking with a *barong* and killing Rodney instantly. A guard leaped from a sentry post as the sailors began to fire their revolvers, and blew the Moro's brains out with a shotgun.

The Mohammedans of the Philippines had originated a unique and deadly method of individual fighting that was a degenerate offshoot of the principle of the *jihad,* or Holy War, that is specified by the Koran. The Sulu Moro, more a Mohammedan in principle than in strict practice, had evolved the rite of running *juramentado* as a means of combatting the Spaniards. Of all of the 200,000,000 adherents of Mohammedanism, only the Sulu Moro practiced this rite which was in direct defiance of the laws of the Koran.

The word *juramentado,* meaning oath, was first used during the final occupation of the Moro capital of Jolo by the forces of General Malcampo in the year 1876. It designated a rabid, wild-running individual who unsheathed *barong* or kris and waged a personal Holy War.

According to the Moro belief, it was within the power of one man, and his kris, to break in a stride from the miserable *nipa* shacks of the Sulu shores to the scented gardens of Paradise where the houris waited. For the Koran offers great reward for the slain in battle. "On couches with linings of brocade shall they recline, and the fruit of the two gardens shall be within easy reach. Therein shall be the damsels with retiring glances whom no man hath touched before them. Theirs shall be the houris, with large, dark eyes, like pearls hidden in their shells, in recompense of their labors past."

With these rewards before their eyes, the young Moros met in the darkness of night, in the mosques, where the *Imams* made elaborate preparation of the body in order that they might appear before God in the most favorable light. Prayers were offered, and the candidates were formed in a circle to repeat the oath of organization. Hands on the Koran, they intoned, "*Jumanji kami batunan ing kami ini magsabil karna sing tuhan.*" (We covenant with God that we will wage this Holy War, for it is of God.) The young aspirants for martyrdom were then bathed, the nails were trimmed to the quick and the teeth were washed. The eyebrows were shaved until they resembled "a moon two days old." The head was shaved, the scanty beard was plucked, and the waist was encircled with a tight, wide band for strengthening effect. The candidate was clothed in a white robe and crowned with a white turban. The genitals were bound tightly with cords, and the body was bound here and there with cords, tightly, to prevent circulation and loss of blood. A man so prepared was able to remain on his feet although dying from fatal wounds.

To the broad belt at the waist was attached the *anting-anting,* the charm that was to ward off the bullets or blows of the enemy. The edged blade, kris or *barong,* was honed to razor edge and beautified and polished, and the Moro was ready to take up that short bloody path that ended in Paradise. These men were known to the Moros as *Mag-sabils,* taking the *Parang-sabil,* or Road to Paradise. We remember them today as *juramentados.*

Man for man, nothing on the face of this earth can equal a *juramentado* Moro in action. Death was their privilege and their reward, and they were fired to a pitch of fanaticism to which no white man could attain. The testimony of a thousand bloody incidents of the occupation of Sulu demonstrates that they were unstoppable. They died; but almost without exception they clung, even in death, to their objective, and they carried with them in death the object of their attack.

On October 17, 1911, one Moro, armed with a *barong* and a spear, ran the gauntlet of sentries of the 2nd Cavalry, stationed at Lake Seit. As the mad Mohammedan hurtled down the company street, the target of fire from all directions, his eyes were fixed on Sergeant Oswald Homilius as the first object of attack. The Sergeant went down, pierced through with the spear. Riddled then with bullets, the Moro turned methodically to the nearest American soldiers. Racing into point-blank fire, he cut down four American troopers with his *barong* before Lieutenant Coppock was able to deliver the full charge of a shotgun at close range.

A few weeks later, one Moro rushed the guard post at Jolo, advancing without cover across fifty yards of open country. Twisting, darting, the howling fanatic reached close quarters and killed two soldiers with his blade before the weight of revolver and rifle fire dropped him lifeless. When the body was examined the Moro was found to have five breast wounds from Krag rifle bullets and four dumdum .45 pistol bullets in neck and face.

I am indebted to Captain J. A. Tiffany, Philippine Constabulary, for the following graphic account of an attack of *juramentado* Moros at Camp Severs.

DEATH ON THE KRIS

"The camp itself was a large rectangle, completely enclosed with wire. The line of company tents were about ten feet inside the wire on each side. Inside the line of tents were the saddle racks and the picket lines of horses. The fence was seven feet high, with ten wires, making the strands about eight inches apart. Every twenty feet along the top of the fence, was a Dietz lantern with reflector to light up the high grass outside for several yards. The firing trench just inside was banked up and ready for business. In a few seconds after an alarm by the sentries, the men could be out of their tents and ready to meet an attack. We felt secure.

"At sundown, with Captain Purington, I inspected the defenses. We agreed that the men could sleep in perfect security with four sentries posted. No Moro could get through that fence alive. Even if they made a quick mass attack, our men would split them on bayonets while they were entangled in the wire.

"I was about ready to roll in that night when I went outside the tent and sniffed the wind like a horse when a bear is in the bush. Lieutenant Crites and myself were quartered in a tent at the opposite end of the camp from our company. Something was not right. I felt it, but could see nothing. The sentries were alert on four sides. I said nothing to Crites about my uneasy feeling. Perhaps it was that I had been used to being near my men at night. In the jungles and in Lanao we Constabulary officers had been in the habit of bunking down alongside our soldiers and non-coms. Here, in an American army camp, we had army traditions to uphold.

"It was in the night that I came out of a deep sleep feeling that a shot had awakened me. Then there *were* two shots and a cry: 'MOROS . . . MOROS.' Then a whole barrage of shots. I reached for my riot gun. It was gone! So was Lieutenant Crites.

Jungle Patrol

DEATH ON THE KRIS

"Snatching my .45 from beneath my pillow, I tore aside the mosquito-net canopy and ran out of the tent. Dark figures were coming up to the fence on the run. The firing was general.

"Realizing that in my white B. V. D.'s I might be mistaken for a Moro, I jumped back into the tent for my khaki shirt, pulling it on as I ran down the company street. Eight *juramentados* broke from cover and charged the camp. The ten second's delay in recovering my shirt saved my life, for I would have been confronted by six of them with nothing but my .45.

"With drawn pistol I was running down the street to my command. My path lay between the picket line of cavalry horses and the row of tents. A dim figure was running just ahead of me. I supposed it was a soldier on his way to the firing trench. The night was so dark I kept butting into the saddle racks. A big cavalryman charged out of a tent just ahead of me with a riot gun. He poked the gun within a foot of the running figure ahead of me and blasted. The man swerved and stumbled on. 'My God,' I wanted to shout, 'stop shooting at our own men.' Then I brought up suddenly. Powder smoke filled my nostrils and I was looking down the barrel of that same riot gun. The big soldier was about to let go again. Some kind of a squealing voice came out of me: 'Hey . . . it's me . . . it's me' . . . I would never have recognized it as my voice. I ran on; there was no time for palaver. My boys were firing rapidly . . . standing up. That puzzled me. I could see the flashes. And then I heard the familiar clang of a steel blade on a gun barrel as one of my men parried a barong. The Moros were through the fence! My men were hand to hand! I saw Crites as I heard the boom of the riot gun. In the red light a Moro was charging in with *barong* uplifted. Crites dropped him in mid-air.

DEATH ON THE KRIS

"Then all firing ceased as the men went at it in a furious bayonet to *barong* duel that was a fight to the finish. At the nearest cavalry tent a white soldier rolled out under the wall, rifle in hand. Before he could stand up a Moro was upon him. Another soldier crawled out and the Moro leaped to him. My Corporal Batiokan ran up to crush the Moro's skull with a rifle butt. Blood was squirting from two great gashes in the cavalrymen's back. Soldiers came running to carry away the wounded man. Their uniforms were red with blood.

"My own company were giving first aid to wounded men. One of the men was past medical aid. He had been chopped to ribbons, with arms and legs severed and lying apart from his body. Under a dead *juramentado* I found a loaded riot gun. I pulled it out and dropped into the trench with my men. Things had grown very quiet. I had the riot gun now; I felt safer. Out in the *cogon* grass I thought I saw something move in the light of the Dietz lanterns. I covered the dark blot and waited. It was a Moro all right. I pulled the trigger and the gun snapped impotently. I fired again with the same result. Then a third shell missed fire and I had a real case of the jitters. Would I continue to snap shells while that fanatic split my head through the wire? Fortunately for me a cavalryman behind me saw me pulling the trigger without result. His Springfield cracked and the Moro went down. The Springfield slug entered the top of his head and continued on through his body. We found him after the fight. He had been knocked down by a bullet in the neck at the first fire of the sentries. Recovering consciousness he had crawled on to be in at the finish. (A Moro *juramentado* has never been known to change his mind.)

"Seven of the eight *juramentados* who had made the attack had succeeded in getting through the wire in the face of the fire. One lay dead outside the wire and seven were stretched out in the enclosure when morning came and we made inspection. The hospital was lined with terribly wounded men, slashed with *barongs,* and we were forced to kill many of the slashed horses who had been in the path of the charging Moros.

"The *juramentados* who had plunged through the wire in a desperate dive had left skin and clothes on the wire. They were horribly torn from head to foot by the long barbs. They were riddled with bullets, and many had heads bashed in and bayonet stabs. They lay there, with glittering eyeballs and bared black teeth. Their heads were shaven and their eyebrows were a thin line of hair. As we looked into those ghastly, inhuman faces and saw those deadly *barongs* still clutched in their hands, it was too much—even for a soldier.

"As I reflected that there might be months and months of this—with every night a possibility of night attack from *juramentados,* it cracked my nerves more than I cared to admit. It was a jittery business, fighting Moros."

The terror brought by these *juramentados* had reached such a point by the middle of the year 1911 that it was decided to disarm all Moros and put an end to the bearing of edged weapons. On September 8, 1911, an Executive Order became effective:

> "The provisions of the Act are hereby made applicable to all Districts within the Moro Province. It is therefore declared to be unlawful for any person within the Moro Province to acquire, possess or have custody of a rifle, musket, carbine, shotgun, revolver, pistol, or any other deadly weapon from which a bullet may be discharged, etc., or to carry, concealed or otherwise on his person, any bowie knife, dirk, dagger, kris, campilane, barong, spear or any other deadly cutting or thrusting

weapon except tools used exclusively for working purposes and having a blade less than fifteen inches in length, without permission from the Governor of the Province."

In 1911, as attempts were made to disarm the Mohammedans, *cotta* warfare began to flame anew and the *juramentados* redoubled their efforts to get to close grips with the American soldiers. Jolo, the Moro capital, in American hands, was almost under a state of siege. It was under constant attack on the part of individual fanatics. One Moro penetrated the city walls through a drain and killed seven soldiers in the streets of Jolo before he was dropped by volley fire of the troops.

For trading purposes, 100 Moros were allowed within the city wall at one time. They were disarmed and searched at the gates by squads of soldiers, and all guard posts mounted four sentries. With all of these precautions, *juramentados* succeeded in running their crazed course at dreadful, frequent intervals.

It was Colonel Alexander Rodgers of the 6th Cavalry who accomplished by taking advantage of religious prejudice what the bayonets and Krags had been unable to accomplish. Rodgers inaugurated a system of burying all dead *juramentados* in a common grave with the carcasses of slaughtered pigs. The Mohammedan religion forbids contact with pork; and this relatively simple device resulted in the withdrawal of *juramentados* to sections not containing a Rodgers. Other officers took up the principle, adding new refinements to make it additionally unattractive to the Moros. In some sections the Moro *juramentado* was beheaded after death and the head sewn inside the carcass of a pig. And so the rite of running *juramentado*, at least semi-religious in character, ceased to be in Sulu. The last cases of this religious mania occurred in the early decades of the century. The *juramentados* were replaced by the amucks . . . who were simply homicidal maniacs with no religious significance attaching to their acts.

As the *cotta* warfare flared in opposition to the disarming measures taken against the Moros, officers and men of the Constabulary began again to win the coveted Medal of Valor. The scene of battle shifted to Mindanao. There, at Mailag Cotta, on February 13, Lieutenant Oscar Preuss won his medal.

With a small detachment of Moro Constabulary he had penetrated deep into the jungle of Lanao, and had come upon a fortified *cotta* of the Moros. Demanding the surrender of the occupants, Preuss had been greeted with a volley of fire. Preuss seized a pole and, with the assistance of one soldier, battered down the gate of the fort and entered to engage the Moros hand to hand. The action was ferocious in the extreme, with neither side giving or asking quarter.

In that attack on Mailag Cotta on February 13, two other Constabularymen won the Medal of Valor. It had been old Sergeant Malaco, only man in the history of the Philippine Constabulary to win two Medals of Valor, who had aided Preuss in battering down the gateway of the *cotta*. With Preuss on this expedition had been Lieutenant Vernon L. Whitney, that gigantic fighting figure of the Constabulary whose size-14 shoes have left an enduring print in the annals of the conquest of Mindanao and Sulu.

Across 40,000 square miles of jungle this hand-to-hand conflict was waging. It was war without a front . . . a war of individual detachments and it was without quarter.

In December a bandit named Pablo de Castro came briefly to life in the northern islands that had been quiet for months. Corporal Telesforo Endaya of the Batangas Constabulary took a patrol consisting of but two men, and went deep into the wooded gorges near Cangapas after the bandit. He met up with de Castro on a twisted jungle trail in high grass. De Castro was under cover, and he fired at the patrol leader at point-blank range. The shot missed, and the battle resolved itself into a personal duel between Endaya and the bandit. The Constabulary Corporal worked through the high grass to a position less than three rods from the concealed de Castro, and as the bandit raised himself for the finishing shot Endaya drilled him with a snap head shot between the eyes. By order of Colonel Harbord, Endaya was awarded the fifty-seventh Medal of Valor awarded to the corps.

But the flurry in Batangas was but an incident. The north quieted again, and the Constabulary attention became focused on Mindanao, Sulu, and Moros. In December, 1500 Moros fortified the old crater of Bud Dajo. Some were induced by General Pershing to withdraw to their homes. The remainder, led by Jailani, were attacked and killed in a second battle of Bud Dajo, which lasted for five days.

Some of the magnificent youngsters were growing old now. Old not in years, but in that unnatural brilliant pace that made them veterans at twenty-five. Few came from the jungle campaigns unscathed. New faces began to supplant the old.

Furlong was approaching the end of a dramatic career. He had given away now, in field efforts, to Whitney and Tiffany and Crites and Cochrun. He was burned out physically and harassed by charges of his superiors in Manila. It was said that he had used unnecessary brutality in his famous Taraca expedition. Hastily Furlong returned from a short vacation to defend himself. He was indicted, tried, and vindicated, and promoted to a Captaincy.

DEATH ON THE KRIS

On detail as Senior Inspector in Lanao, Furlong demonstrated the old fighting genius that had made him one of the most powerful figures of the Constabulary. But his old vitality was gone, and he was gnawed by thoughts of his trial and the attendant publicity. Always a strange, sensitive figure, he broke at last under the strain of the years of jungle campaign. He was sent to Manila for observation and treatment, arriving there on June 21, 1911.

On the evening prior to Furlong's death he dined with the officers at the mess, and during the meal gave no sign of depression. At nine o'clock in the evening of July 9 he passed two officers on his way to his quarters. A moment later a shot was heard; and when they entered his room, Furlong was found dying on the bed from a gunshot wound.

He was certainly the most romantic, and without question one of the greatest, individual fighting men of that long line of fighting men who pacified the Philippines. In his short span of life he lived a dozen lifetimes. He was typical of an age . . . a desperate fighting age when youngsters grew hurriedly to full man size. His memory will live forever in the archives of the corps he served so well.

Chapter xviii

THE SECOND DECADE

"In disturbed districts a number of patrols may be in the field at the same time. Extreme care must be taken to prevent mistakes in firing upon each other . . ."

—*Constabulary Manual*

THE Constabulary, though unaware of the fact, was now entering into the final phases of the active Moro resistance. Under the hammering of the patrols the Moros were breaking up into small detachments, and the conflict was degenerating into disorganized guerilla warfare. But before that day when warfare was to become completely unofficial, there remained several sanguinary battles that were to establish definitely the prestige of the Insular Police.

The battle at Sahipas Cotta in January, 1913, was one of the fiercest combats of the Moro wars. Near Taglibi, twelve miles from the city of Jolo, the Moros had built up a solid rock *cotta* and prepared for its defense. As usual, the *cotta* was constructed on fairly level ground, with stretches of *cogon* grass on all sides. The walls were of huge boulders rolled together to a height of ten feet, and pierced with loopholes for the concealed riflemen. Under the walls were firing pits to protect the defenders from shrapnel fire. Just outside the fort, and circling it, was a loose fence of bamboo, open in structure to offer no impediment to rifle shooting, but strong enough to prevent attackers getting through without delay.

Against this fort a force of Scouts under Captain McNally and Lieutenants Townsend and Cody, and a Constabulary detail commanded by Lieutenants Cochrun, Whitney, and Crites went into action to open the year 1913.

For some hours the attacking force lay under cover, while mountain artillery thoroughly shelled the Moro position. After bombardment by one-pounder shrapnel shells the shacks inside the fort took fire and McNally, an impatient fighting Irishman, ordered a charge.

The line of Scouts rose from their positions and rushed the walls. But the Moros, secure in their rifle pits, had been untouched by the shrapnel, and as the Scouts boiled up to the bamboo fence, hacking and slashing with bolos at the impediment, the walls of *cotta* flamed with rifle fire. McNally took a slug in the body, and went down at the edge of the walls. Five Scout soldiers fell beside him, and nine Scouts were seriously wounded. Without making contact with the enemy, the Scouts were forced to withdraw, leaving their dead and wounded under the *cotta* wall.

Then, from the flank of the Moro *cotta* came the charge of the Constabulary, led by Cochrun and Crites. The lanky Cochrun, six feet in height and towering above his native soldiers, made a perfect target, and as the 6th Company of Constabulary came under the walls, he received a slug through the right forearm that shattered the bones and rendered the arm useless. Transferring his Colt .45 to his left hand, he moved ahead and received another bullet in the stomach that dropped him in his tracks. Two soldiers at his side died, shot through the head, and along the line the Constabulary wavered as a half dozen men went down wounded. The Moros in this battle exhibited the finest marksmanship of any battle of the wars.

Meanwhile, Whitney was clawing over the fence when a bullet passed through his left leg, severing the sciatic nerve. At his side, Lieutenant Townsend of the Scouts went down with a bullet through his side. The Moros were making sure of the officers. Only Crites of the Constabulary and

Cody of the Scouts were on their feet, and the young junior officer of Constabulary took command and ordered a withdrawal. In the face of the direct fire from the walls, the dead and wounded were gathered up and the force withdrew to shelter.

For two hours the battered American force gave first aid to the wounded, listening meanwhile to the shouts of derision of the Moros, who were maintaining a steady fire. Then the one-pounder began to pound away at the *cotta* wall. The shrapnel ripped through the bamboo fence that encircled the fort, firing the weckage. Then the little cannon shifted to high explosive shells and began to batter away at the rock wall. The wall began to crumble.

Late in the afternoon, the two Lieutenants began to organize for another charge. About the fort was a ring of live coals from the burning bamboo. The Constabulary soldiers could face rifle fire, but they found they could not cross the barrier of coals in their bare feet. One squad, led by a private, did leap across the flaming barrier and into the *cotta,* mopping up the defenders on one side. The Moros reorganized and drove them from the fort, and once again the American force withdrew, carrying with them their wounded.

Camp Taglibi was three miles distant, and during the night the dead and wounded were evacuated thither for embarkation on the *Jewel* to the hospital at Jolo. After a brief rest and breakfast, the troops returned to the *cotta,* prepared to renew the attack. But the Moros had evacuated the shattered fortress in the night and moved to the strong position atop Mount Bagsak, in the crater of an extinct volcano.

For this engagement, First Lieutenant James L. Cochrun, 6th Moro Company, was awarded the Medal of Valor.

Some months after the assault on Sahipas Cotta Lieutenant Ellsey of the Constabulary was sent into the hills to serve a warrant on a Moro named Usap for stealing carabao. He had anticipated no particular trouble, and carried with him a small patrol of six men.

He found his man standing in the door of the usual Moro shack, with a ladder leading up to the door. The Moro glowered down at the small patrol as Ellsey served his warrant. His expression did not change as he turned to get his turban for the trip. But Ellsey felt that all was not well. He circled the shack and saw Usap reach under a mat and draw forth a *barong*.

The Constabulary Lieutenant raised his rifle and drilled the Moro through his head. As Usap dropped, two other Moros leaped from the room. The waiting patrol dropped them in mid-air. They were dead when they hit the ground. The patrol then mounted the ladder and captured three additional Moros who had not yet worked themselves into the amuck stage.

While they were tying these prisoners beneath the house, a Moro in a near-by field was plowing rice with a *carabao*. They heard him shout as he leaped to attack with a *barong*. "*Timbuck aco,*" he was shouting; "shoot me." He came with long bounding strides, headed straight for the waiting patrol.

Four of the soldiers opened fire on the advancing Moro in support of Lieutenant Ellsey. A stream of hot lead poured into his body, but the Moro never faltered. He came nearer, slower now, but still on his feet. The *barong* was upraised as he headed for Lieutenant Ellsey. Ellsey fired his last shot, and the Moro still came.

Ten feet from the officer a Krag bullet thudded into the amuck's spine. His legs gave away. As he fell, he hurled his *barong* before he died.

The patrol stripped the dead man and turned him over.

Twelve bullet holes were in his body. Ellsey had escaped decapitation by only ten feet.

The patrols spread out to cover more fully the Moro Province; Lieutenant Polk operated from Guthries Post at Tubid Cove. With Captain Fawcett and Lieutenant Hoffman on station at Capay, two strong detachments were in the field in the vicinity. Captain Fort and Lieutenant Johnson were with the General Service Company, trouble-shooting through the hills. Tiffany was constantly in the field searching for outlaw bands.

Tiffany made the capture of old Amai-gin-dalungan his special task. All of May and June of 1913 he was on almost constant patrol in the territory of the wily old outlaw. In collaboration with Captain Fort, he had several brushes with the Moro, but Amai always managed to elude the patrols.

In July, Fort, Johnson, Polk, and Tiffany converged upon an outlaw clearing on the shoulder of mountain at a place called Gaba. As Moros began leaping from the shacks, the Constabulary opened fire. The range was 400 yards and it required fast, accurate shooting. Tiffany lined his sights on a Moro in the act of leaping. It was a hit, and since their main objective was to capture one of Amai's men, the patrol hastened to the fallen Moro. They followed a trail of blood and pulled the Mohammedan from beneath a log. The soft-nosed slug had torn a hole in the right lung the size of a quarter. Polk and Tiffany packed the wound and questioned the wounded Moro.

From him they learned the location of the band of Amai. Tiffany set his men to work on the wounded Moro, whose name was Pakalangut. By morning they had convinced him that the life of an outlaw was very foolish. They shared betel nut with him, and he announced himself ready to lead them to Amai's camp.

But Pakalangut, with a hole the size of a quarter in his lung, was very weak. After half a day's march into the interior, he fainted, and the advance halted. Tiffany revived him with brandy and the advance moved on. At the base of a rugged ravine the guide halted and whispered that they were within a quarter of a mile of the outlaw outpost. Polk and Tiffany divided their force into two detachments, each taking ten men.

Tiffany worked his way hand over hand up the mountain side when the sound of Polk firing opened the attack. The Moros broke from their camp in surprise and took to the brush. Tiffany saw two Moros helping along a third, who walked with difficulty. His Krag bullets followed them into the brush. Hastening forward, he found blood on the ground but no dead Moros, and a moment later the two detachments united with no apparent result from their attack.

Not until the following August did Tiffany learn that his volley into the group of three Moros had fatally wounded Amai. The old Moro chief had lived but a few hours and had been buried in the jungle.

More fortunate was Pakalangut, the guide with the jagged hole in his lung. He lived to become a valuable Constabulary guide at Dansalan.

In accounting for Amai-gin-dalungan, Tiffany accomplished a feat that had been the goal of all Constabulary officers for a period of half a decade. Amai was shrewd and powerful and a great leader. His death robbed the Moros of capable leadership in Lanao.

Tiffany must be ranked among the first files of the superior fighting men who made up the Constabulary roster. His record of continuous Moro service was excelled by few of his fellow officers. He had special qualities fitting him for leadership of the Moros and he was greatly respected by his wild, Mohammedan soldiers. He took over a Moro company that was riddled with dissatisfaction and desertion, and made it into one of the finest companies in Sulu.

When Tiffany finally resigned to enter the World War, the company passed to the command of Lieutenants Shamotulski and Wade. Wade and eight of his men were ambushed and slain by a rebellious *Datu* who had been friendly to Tiffany.

It was Tiffany who believed he had solved the mystery of the Moro means of rapid communication. His theory was that the robber bands used kites as means of spreading news of the approach of a patrol. The Moro kites were about five feet in height and had attached a rattan bow, with a deerhide thong. When the kite was high in the air, the thong vibrated with a sound that could be heard for miles. In Tiffany's opinion, the various colored tails used on these kites, were the code messages that signaled, "All clear," or "Patrol approaching."

It is also probable that the Moros used the drums fashioned from a hollow hardwood log as a means of communication.

Second Lieutenant Samuel T. Polk, who had operated with Tiffany after Amai, had won his Medal of Valor a few months previously on the Labangan River in Mindanao. Trailing the Moro outlaw Mapandi, Polk had ordered an advance on the Moro position, and in the ensuing fight a private of his detachment had fallen in an exposed position in front of the outlaws, shot through both legs. Polk had moved up to within a few yards of the bandits, who were concealed behind boulders, and had carried out the wounded private under direct fire.

Across a valley and a mountain range, Lieutenant Donald Root, a restless young New Yorker, was leading a patrol against the band of Raja Muda Randi in the vicinity of Mamaya Peak in Lanao. The mixed command of Constabu lary and Scouts had departed from Tamparan, Lanao, on December 13, 1913, and at two o'clock in the afternoon on the fifteenth, the column encountered Randi's band in dense jungle on Mamaya Peak.

At the first volley of the outlaws Root received a serious gunshot wound in the side, but he continued to advance until, weakened by loss of blood, he was forced to halt. Supporting himself against a tree, he continued to fire and direct his men until he fell unconscious. He won his Medal of Valor at the age of 21, the youngest officer to achieve that distinction.

Minor engagements were of daily occurrence during these preliminaries to the pitched battle at Bud Bagsak. Lieutenants Gunn and Gilmore were in the Sarangani Bay region of southern Cotobato. Youngblood was on the shores of Lake Buluan, against Moros and Bilaans. Tiffany, Tarbell, Preuss, and Wood were in northern Cotobato. The assault of each *cotta* paid a price in lives. From 1901 until June 30, 1911, the Constabulary had patrolled more than 10,000,000 miles, fought in more than 10,000 engagements, and suffered heavily in casualties. The records show officers killed, wounded and dead to the number of 104 in that ten-year period. Enlisted men killed, wounded, and dead totaled 1,602.

The Moros were standing fast to a principle that was centuries old before America came to the Islands—their inherent right to bear arms.

The assault on the five *cottas* of Pujacabo, Bunga, Matunkup, Languasan, and Pujagan was not a Constabulary engagement. General Pershing, in his capacity as commander of the armed forces in Mindanao and Sulu, ordered out Scouts and regulars after it was seen that arbitration was useless. In a terrific five-day battle that saw mountain artillery, Scouts and infantry in action, the last great organized resistance of the Moros was waged. Five hundred Moros occupied the *cottas* on the mountain top, and at the close of the campaign they were dead to the man.

The Constabulary had only a minor part in the battle, which was of the close order, massed attack, suitable to regular army tactics.

To Captain Jesse C. Tiffany I am indebted for this Moro version of the battle of Bud Bagsak. Better than any American account of the battle, it expresses completely the attitude of the Mohammedans:

"When America came and the Spanish devils were driven out of Sulu we rejoiced. I saw your Honorable Taft in Jolo. He smiled at our Sultan and waved to the people. These Americans are friends of the *tao Suug* (Sulu Moros), we thought. When your Generals, Bates, Scott, and Wood went among the people and made friendly talk with our *Datus* we were sure of it.

"But America changes face. One man promises and goes away. Another follows and will not perform.

"Soldiers came and the Constabulary. For us you made laws without our consent. Always before we had slaves to do the work and a man could have two or three wives. But now we must give up our slaves and have only one wife. Our *Datus* were pushed aside. New Judges came from Manila. No friend of the Moros came from Manila. Our laws were ignored. Moros were judged by laws we knew nothing of and were sent away to jail. They did not come back. But their message came to the people. Prison and jail were worse than death.

"Somehow I knew. The Panditas were false leaders. We could not fight your soldiers. Their guns would kill as far as you could see a man. Their cannon could blast down a mountain. What could we do with *barongs,* and a few *sinapans* (guns) that we had smuggled in from Borneo? There was no hope in our arms. There was no hope, save Allah.

"As you Americans say, 'we had a rendezvous with death' on Bagsak. We were not to be denied. There was no reason in the hot-heads. Better to give battle now. In a few months all our guns would be gone. And next they would be asking for our *barongs*—which you are doing now.

"For months we had been making our *cottas* on Bud Bagsak secure. There were five of them, and our main *cotta,* Bagsak, had thick walls of stone. From deep firing trenches and pits, our bamboo tubes commanded the slopes of the mountain. One of our rifles could stop a hundred soldiers on the steep slope of the mountain, so we believed.

"But the Americans were hard to hit. They hid in the grass and brush and climbed the rocks like cats. And they could shoot straight from a long distance.

"Before they got up close enough for us to shoot them their shells began to burst in four of our *cottas,* Matunkup, Languasan, Bunga, and Pujacabao, killing many of our people.

"We were in the main *cotta* of Bagsak, my father and I. But my uncle and brothers were in Matunkup. Toward noon my brother came up to us horribly torn, deaf, and almost blind. The jagged metal from shells had torn the flesh of his legs to shreds. He said the inside of the fort of Matunkup was roaring with exploding shells like a volcano. There was no air to breathe. Only dust and acrid powder smoke. My poor brother wanted to die but couldn't.

"Then came more wounded. In the face of exploding

shells they could not see to shoot. An American officer with his soldiers climbed the face of a cliff and took Matunkup. The mountain shook and seemed aflame with guns. The roar of the cannon was like heavy thunder. In a few hours they pounded Pujacabao to pieces. Amil, our leader, came up wounded. He foamed at the mouth and screamed for Allah to send fire and brimstone to avenge the Moros. Before night Languasan had also fallen.

"That night we rested on our arms. My brave father wanted to lead a hundred *barong*-men into the American trenches at night. He would drive the soldiers off the mountain or die. But Amil would not give the order. 'Wait,' he said. 'There is tomorrow and there is Allah. When the battle turns we will mow them down like grass. None shall escape.'

"All night we worked constructing shelters to shield us from exploding shells. The next day was ours. We held our own. Our bullets kept the Americans crawling in the grass. . All day their mountain guns pounded *cotta* Bunga. But our men fired from bomb-proof pits beneath the walls. We beat off every attack. Foolishly, Amil ordered counterattacks. He fell, likewise several *Datus,* and all the Moros who charged the Americans. None came back.

The next day the fight raged around *cotta* Bunga. After hours of slaughter on both sides, Americans climbed the walls. Inside was carnage, as the Moro likes it. *Barong* against bayonet. The defenders died to a man, but their blades found the enemy. My father was wounded that day.

"There remained Pujagan and Bagsak, our last stand at the rim of the crater. After a hard day's fighting, the Americans seemed discouraged. Our leaders predicted they would give up the fight. The cost of taking Bunga was more than they had the bowels to pay. Another fight and we would go forth to the slaughter.

"But another night never came. It was Sunday, the Christians' Sabbath. Maybe that was why. Anyway, it was not our day. Allah and Mohammed seemed to be hiding behind a sky of brass that day. The Americans had dug trenches close. From these came a stream of bullets. Our men in the trenches outside the walls of Bagsak traded with them bullet for bullet. Then came a hail of shells that men above ground could not endure. They pounded our outer defenses to bits. The few men living had to retreat to the *cotta*. All day they pounded us with shells inside the *cotta*. Only a few of us lived through it. The walls were finally crumbling like clay. Firing pits caved in to bury my father, mother, and sisters alive. My own life seemed charmed. We fought on doggedly, looking to the only escape—death. Several rifles and muskets burned out in my hands.

"Now the Americans were not three lengths of a coconut palm from us, firing from our old pits. Their cannon sprayed the top and inside of our fort constantly with metal so that we could not show our heads to shoot down on them. Yet our fire was so hot they dared not spring from their trenches for the final rush. Their fierce attack was matched by our own desperate resistance. Thus we were deadlocked while the sun climbed across the brass dome of heaven and dropped down toward the rim of the world.

"Suddenly the cannon ceased. We rushed from our holes to man the top and meet them with our *barongs*. Horrible creatures we were, red eyes rimmed with clay, naked, blood-caked bodies, wounded, bleeding, crawling. My wounds were so many I no longer felt them. I managed to stand on top of the wall of Bagsak and fire into the faces of the soldiers. I saw a man with red hair and fierce glaring blue eyes topple back and fall as my shot caught him in the stomach. Another soldier as he vaulted to the top of the wall was slashed in two pieces by one of our *barong*-men. I drew my *barong* and slashed at the head of another soldier, and then things turned black. I was falling."

Following the assault of Bagsak, the back of the resistance was broken, and the Moros were split again into small guerrilla detachments. The army withdrew, and the patrol duty in the interior was resumed by the Constabulary.

Driven from Jolo, the worst of the bandit element took refuge on Basilan Island, across the straits from Zamboanga. It was there that Lieutenant Ernest Johnson met death while pursuing the band of Moro Atal. The patrol, with Johnson at the head, was moving along a narrow jungle trail, to walk squarely into ambush. The first warning was the whine of a thrown spear and Johnson went down, pinned through both lungs. Lieutenant Ernest Johnson was a graduate of the University of Nebraska and the University of Oregon. He was greatly respected by his fellow officers for his fighting qualities and unfailing good disposition.

Chapter xix

THE LAST OF THE COTTAS

"Sentries may frequently be more useful by standing and watching what goes on than by walking their posts . . ."

—*Constabulary Manual*

IN 1913 the old guard members of the Philippine Constabulary—the bush veterans who had conquered jungle and made the country safe for American citizens—reaped the usual reward for such service. With the arrival of Governor-General Francis Burton Harrison, the Constabulary was the sole surviving monument of the old traditions. Under Harrison, a system of too rapid Filipinization took place, and members of the Philippine government service found themselves ejected in middle age from the army and civil lists of the Philippines.

Out of touch with their homeland and weakened by years of malaria and jungle warfare, the old campaigners found themselves the victims of a cleverly worded law that forced their retirement. This law provided that every American official and government employee who applied for retirement within three months could retire with a bonus of one year's pay. The result was eviction by wholesale of Constabulary officers.

But in the southern islands, a nucleus of old veterans remained to carry on the bloody work of the pacification of the Moros. They were men apart in a materialistic world; they fought for the joy of fighting, with an ideal of service before their eyes and a vision of the Philippines that were to be.

And to those Constabulary riflemen of two decades ago, the rubber planters and plantation men of Mindanao owe their existence today.

Captain Vernon Whitney became Governor of Jolo. In assuming the position, Whitney was undertaking the leadership of a jungle province that was the oldest battleground in the world. Men of every creed and color had tried to occupy the city of Jolo and its environs. Few had succeeded in impressing governmental discipline upon the Moros. Esteban Rodriguez de Figuoea had gestured feebly with the Moros in 1578 before falling to a kris blade in Mindanao. Behind him, great Spanish names had appeared, briefly to gesture with death. Shadowy figures for the most part, and deep in the finality of oblivion: de Sande, the religious fanatic; Juan del Campo, Gaspar Gomez; Juan de la Jara, the amorous; and Captain Paches, a magnificent figure for an hour, before he relaxed in death on the outpost walls at Caldera.

History yields the names of the Moro campaigners reluctantly. Lorenzo de Olaso was one . . . a glittering figure in armor, advancing too near the walls of Jolo to go down before a whining spear . . . Cepeda, leaving 400 Moro dead on a reddened field, falling in turn at a bloody ambush on Tawi-Tawi . . . . Valiant Spaniards, feeding their bodies to the Moros but replaced and replaced by others as valiant.

This Moro capital of Jolo had been assaulted by Spain on sixteen occasions, five of which resulted in the occupation of the city. It required Spain 298 years to establish a civil government within the walls of Jolo; and of the 321 years of struggle for the city, it was held by the Moros for 290 years and by a Spanish garrison for 31 years. Whitney was taking over a position that was full man-size.

After the battles of Bud Bagsak and Talipao, the Moros

were of uncertain temper. There was no withdrawing from the struggle; if the Mohammedan official army was broken, the mood for battle was still strong in the air. The patrols reported that often the mere sight of Constabulary bayonets was sufficient to start amucks on their course of death.

Shortly after Whitney became Governor, reports began to sift into headquarters of trouble between two Moro *Datus* at Bual on the eastern end of the island. Whitney was no swivel-chair Governor. Hastily he ordered out the launch *Jewel* and proceeded to Bual. He found the Moro market-place a scene of apparent peace and quiet. His interpreter, a Moro named Arolas, questioned the people for a few moments and then asked permission to visit friends on the beach.

The Moro did not return, and Whitney set off down a coconut palm shaded trail in search of him. The Governor was moving slowly, as one leg was almost lifeless from his severe wound at Sahipas Cotta.

Nearing the beach, he met two Moros in the trail who approached him with the information that his interpreter had been wounded and was in need of attention. Then they were upon the Governor with bared bolos.

But Whitney, giant in size and magnificent in personal combat, acted with the instinct of the born fighting man. As the Moros rushed him he reached for one and with a simultaneous motion, tucked the Moro under his arm with the left hand while his right drew a .45 and blew out the brains of the other. Whitney then placed the muzzle of his revolver to the ear of the other Moro and the affair was finished in five seconds.

But the Governor was not unscathed. While Whitney was firing at the second Moro, the Mohammedan he held captive under his left arm had squirmed an arm free and laid open Whitney's back from shoulder to spine. A month in the hospital and he was on the job again . . . with a prestige of value as a fighting man that was to win him the highest measure of success of any governor of Jolo of those early days.

The disarming of the Moros was splendid in theory, but almost impossible to execute without great blood-shed. The Moro had borne arms for centuries, in a land where the bearing of an edged blade was the sign of manhood. It was a privilege that his Sultans and *Datus* had encouraged; and to a Moro, the guarding of a privilege is a jealous rite.

The Constabulary was under orders to take every *barong, campilane* or kris they encountered in the course of a patrol. It was desperate, bloody work. In speaking of this law Lieutenant Tiffany writes in detail:

"It was a terrible thing to take the *barong* away from a Joloano Moro. You were taking away his visible masculine characteristic. You made him a woman and less than a woman. Most any Constabulary officer could kill a Moro and take his blade. Some officers did. It was all a part of the day's work to them. When they met a Moro wearing a *barong* they called for the blade. If he resisted or started to run, they shot him and entered it in their report. But to take a weapon from a Moro required skill and patience, and I could not find it in me to kill them in cold blood because they stood on their tribal rights.

"To reverse the situation: if Mexicans swarmed across the border and started killing every American who refused to give up his gun we would call them barbarians and fight them to death with teeth and pitchforks. So I had sympathy with the Moros and sought to make it easy for them. But I had my orders.

"I would be leading a patrol across Jolo, with perhaps six to ten soldiers. We would see a Moro cutting through a field and notice that he carried a blade. The carved mahogany handle would tell us it was a fighting weapon. I would hail him.

"He would approach and I would try to remonstrate with him. I would tell him that the world was full of men—real men—who did not carry *barongs*. That it was against the law and he must give it up. Sometimes he did. Sometimes he would approach, and when he would see the soldiers his eyes would begin to glisten with excitement. When I saw that he was going to make a run for it, I would order the men to fix bayonets. The sight of the blades and the rattle of the bayonet as it went on the barrel usually tripped the poor fellow off. He would whirl and charge, deciding that life without a *barong* was not worth living.

"The Moro soldiers would not be sorry for him as they shot him down. They knew that he wanted to die. Sometimes he would cast himself upon the bayonets. Being a Moro, he preferred to die on a blade. Yes, disarming Moros was a sad and messy business."

The mere killing of Moros was not to solve the problem of Sulu. The highest type officers in the Constabulary—men like Tiffany and Whitney and Fort and Wood,—were quick to see that. Nor was the Moro to be bluffed. The work of the Constabulary was developing into bona fide police work. The old time fighting man and combat officer were giving away now, whenever possible, to the diplomat and advisor, but there was no love for Americans in Jolo. The mountain-top battle of Bud Bagsak was the high tension spark that set the whole island aflame. The Moro recovers quickly from a defeat in the field. War is a game, and a worthy antagonist makes the game better worth the playing. And these Americans were worthy antagonists. Here were no soldiers cowering behind walls of stone . . . here were young fighting men who sought the open fields of *cogon* grass to meet the Moros on their own ground.

The result was a reversal of the modes of warfare in the Moro country. The Mohammedans dug in—the Americans went in after them. All over the country *cottas* began to spring into being, thick-walled forts of packed earth and rock, defended by firing pits and slashing krismen. Sometimes the Constabulary patrols were able to surprise the natives in the act of constructing a *cotta*. A few sticks of dynamite and the fort would go up in ruins . . . a week later and a passing patrol would see a new fort erected on the ruins of the old.

Shortly after Bagsak, two *cottas* sprang up overnight, almost, in a ravine near Jolo. The 16th and 24th Scout Companies and a mountain battery under Lieutenant Dillman were ordered out to raze the forts. The Scouts were under command of Lieutenants Walker and Conroy.

For hours the Scouts pumped shrapnel and solid shot into the *cotta* of Tahil. The Moros on the walls laughed as the shells bit into the walls. Dillman was getting no results from the cannon fire other than a strengthening of the bamboo walls. A squad of dynamite bombers was ordered into action, and the walls began to collapse. Handling the dynamite bombs was a tricky business. There was always a chance that a Moro bullet would set off the caps, and the fuses were cut very short to prevent the Moros from throwing them back on the troops.

As the wall collapsed the Scouts went over the top and bashed in the Moro's heads with rifle butts as they emerged from their firing pits.

A hundred yards away the *cotta* of Jahanal was still under attack. The Scouts had to dig their way into the fort while a squad of riflemen kept the walls clear of Moros. When an entrance had been made in the fort walls the Scouts were unable to capture the fort, as the Moros were in position to lop off the heads of the attackers as they crawled through the small opening. Walker worked a squad of men and a mountain gun against the side of the fort and pumped shrapnel at point-blank range into the *cotta*. The Moros were ground to bloody powder, and the Scouts entered to finish off the remainder in hand-to-hand battle.

In the whole history of the Mohammedans there is no evidence of unwillingness on their part to die in battle. As the disarming process continued, the hot-heads banded together again, and when the Star and Crescent hung low on the horizon above the Southern Cross the signs were right for another resistance to American authority.

It was on the summit of Bud Talipao, near Maimbun, that the Moros gathered for another of their last stands. Talipao was to be the Mountain of Destiny, and through all the southern islands swept the word that Allah was standing by to deliver a miracle. Rice thrown into the air would turn into bees that would swarm down the mountainside to sting the Americans. While the Americans were fighting the bees, the bolomen would rush them and Sulu would be delivered.

The assault on Mount Talipao was begun on October 24, 1913. Major Shaw of the Scouts led the attack, which got under way early in the morning of a day that was still and sultry and uneasy. The companies deployed through the high *cogon* and moved up and over the first rise of ground. Not a Moro was in sight. Not a shot was fired until the troops had moved 200 yards up the slope.

Then the Moros came. With a yell, they unsheathed the terrible *barongs*, and the mountain side was colored with the bounding figures dressed in green and blue jackets and orange and red turbans. Fifty Mohammedans made up that first assault, and the line stood firm to receive them, They fell, riddled with bullets, without inflicting casualties. The Scouts pushed up the mountain. Another fold in the mountain disgorged half a hundred fanatics. The rifles grew hot as the Scouts faced charge after charge of frantic krismen. It was desperate work, with death for the soldier who missed the fast moving targets.

At the summit of the mountain the troops came out to see the rifle pits of the defenders and the main body of the Mohammedans. As they fixed bayonets for the last charge, the Moros were outlined against the sky, throwing rice into the air. The Moro leader, a fine individual marksman, seized a Krag rifle and opened fire on the Scout line. Six bullets from that deadly rifle found the hearts of Scout soldiers.

The advance halted. Major Shaw, leader of the attacking force, saw that the eviction of the Moros as the result of a charge across open country would cost enormously in life. He ordered a squad of riflemen to maintain a hot fire across the pit where the concealed *Datu* marksman was hidden. This enabled Lieutenant Conroy to dash close to the trench and throw in an improvised dynamite bomb.

But the Moro was ready for him. There was a blur against the sky as the bomb came back, to explode in the faces of the attackers. Conroy threw another bomb, fuse shorter this time. Again the bomb came back, hurled by the Moro, to explode in the air. Then Conroy, the bomber, prepared a bomb with a fuse that was timed to cause an explosion in five seconds. Signaling the troops to pour in a terrific volley, Conroy sprinted to within thirty feet of the rifle pit, touched a match to the fuse and tossed the bomb to the waiting Moro.

KATIPUNAN SYMBOL OF THE FILIPINO TRIBES
(Copy of Drawing in the Original 1938 Edition)
See Chapter Five - The "Popes"

BATTLE FLAG OF THE MOROS
(Copy of Drawing in the Original 1938 Edition)
Flag had a red ground, with blue canton, and white stars and weapons.

Sultan of Sulu with Kris Guard

Representative Moro Warriors

In that manner ended the third decisive victory of the Americans over the Moros in Sulu. But that exploding bomb that sent a Moro patriot to Paradise did nothing to quiet the mood for battle that remained.

Mindanao was swarming with malcontent Moros who were at grips with the patrols in all parts of the provinces of Lanao and Cotobato. *Juramentados* with shaven heads were running the jungles of Basilan, less than fifteen miles from the American headquarters at Zamboanga. In the city of Jolo, *juramentados* were penetrating the town almost at will. The records of the day tell of the visit of one *juramentado* who penetrated the walls and killed nine persons before he was blasted out of existence by combined fire of the guard post.

But there was another side to the embattled Moros—a side that is rarely considered in any estimate of these Mohammedans. That was their undoubted ability as businessmen. The Sulu Moro has had the benefit of many centuries of traffic with the white man, and he has brought to these sessions a native craft that is the birthright of the Malay. As a fighting man or a financier, the Moro must take a high rank.

The establishment in 1936 of a tentative Philippine Commonwealth Government to replace the American colonial experiment brings to light certain financial transactions remaining to be settled by the new régime in the Philippines. They concern the emoluments of that Royal Financier, the Sultan of Sulu.

The Sultanate of Sulu came into being in the year 1450 with the accession of one Sayid Abu Bakr, a recent and persuasive arrival from Borneo. Abu and thirteen of his successors were individual entrepreneurs, dependent upon the bounty of the country and the efforts of the pirate squadrons in that matter of filling the royal treasury. They achieved a certain measure of success, but with far greater effort than was necessary in the later years of the dynasty.

It was not until the crowning of Alimud Din I, about the year 1735, that the full benefits of civilization became apparent to the Sultans of Sulu. True, during the reign of the sixth Sultan, he who had been called Mohammedul Halim, the Spaniards had arrived in force before the capital city of Sug (now called Jolo), and opened negotiations for contact. The Spanish commander, Don Estevan Rodriguez de Figuroea, bore with him a letter from the Spanish Captain-General, Francisco De Sande, containing instructions as follows:

> "You shall repair to the Island of Sulu and you shall there bargain with the natives as to what tribute they shall pay, which shall be in pearls . . . you shall take from them only what is necessary for food and the provisioning of your ships for the return voyage . . . if the natives of this place shall give tribute, you shall act according to the usual custom — namely, you shall place one-half to the account of our Majesty, whilst the other half shall be distributed among the soldiers."

The Sulu Moros were singularly irresponsive to this beneficient Spanish suggestion, and the idea was regretfully abandoned by Figuroea.

This then, had been the extent of the Spanish contact, with the exception of desultory conflict, until the year 1746, when the Sultan Alimud Din I became suddenly aware of the benefits of outside financial assistance. He granted, therefore, the admission of Jesuit missionaries to Jolo, and wrote to the Spanish Captain-General with a request for 6,000 pesos, 1,600 pounds of gunpowder, and 1600 pounds of nails.

The nails were for the construction of suitable quarters

for the priests of Spain—this in a country where the houses are built of *nipa* palm, lashed together with vines! The need for the gunpowder remained unexplained in the Sultan's letter. Sometime later, when Mu'izzud Din II had usurped the Sultanate, the need for the gunpowder became increasingly plain to the Spaniards, for the year 1753 was the bloodiest in the whole history of Moro piracy. The nails were used with telling effect, and the *conquistadores* spent weary hours picking them from various portions of their anatomy. They had made splendid charges for the brass cannon of the Moros.

The next monetary concession of importance to be made by the Spanish government to the Sultan of Sulu, occurred in the year 1836, coincident with the signing of a commercial treaty prepared by Captain Jose M. Halcon. Its provisions related to the payment of port duties by Spanish craft anchoring in the Sulu capital at Jolo:

Article 5—Spanish craft in Jolo will pay the following duties:

	Pesos
Ships of three masts from Manila, with Chinese passengers	2,000
The same, without passengers	1,800
Brigantine from Manila, with Chinese passengers	1,500
The same, without passengers	1,300
Schooner from Manila, with Chinese passengers	1,400
The same, without passengers	1,200
Pontin (small trading boat) from Manila with Chinese passengers	1,400
The same, without passengers	1,200

It may be seen from the above that the Spanish foothold in Sulu was insecure, and very financial in nature.

By the year 1851, the Sultans of Sulu were demanding

a place of dignity on the Spanish official payroll. It was made plain to the Spanish authorities that the Moro pirates had been singularly inactive, and that an outbreak might be expected at any moment. The Spaniards took the suggestion kindly and a new treaty came into being, signed and sealed on April 30 of that same year. After some haggling, the amount of the Sultan's salary was established at 1500 pesos per annum. In addition, he received an "adequate Royal Title" from the Spanish King.

Four years later Sulu pirates burned the town of Zamboanga.

Affairs drifted along in desultory manner for several years, but in 1878 it was decided that the salary of the Sultan was not sufficient to guarantee order, and another treaty was signed in July of that year. The new deal provided for an annual salary of 2,400 pesos for the Sultan, and it was signed: I, Don Domingo Moriones y Murillo, in the name of his Majesty the King of Spain, Alfonso XII, whom God keep, do confirm and ratify the above Act of Pacification and Capitulation, in all parts."

Even that was not sufficient to secure a lasting peace treaty, and the Moro wars continued on without interruption until that day in May, 1899, when the Spanish garrison lined the pier to watch the landing of the first American troopers under Captain Pratt.

The dollar displaced the peso in Sulu. Brigadier-General J. C. Bates now undertook the arrangement of the Sultan's financial subsidy in the name of the Government of the United States. The *Bates Agreement* of August 20, 1899, was the result of the negotiations, and it provided an annual retainer to the Sultan of three thousand American dollars. The Sultan thus found the change in paymasters most beneficial.

It should be mentioned also that His Majesty, in his other capacity of Sultan of North Borneo, was in receipt of a substantial payment from the *British Borneo Trading Company*, thereby acquiring a familiarity with currencies of two great nations.

This pleasant association with the American eagle, as pictured on American currency, was continued until March 21, 1904, at which time the Sultan visited amiably with Governor William H. Taft in Manila and succeeded in having his financial remuneration increased to the sum of $6,750 each twelvemonth.

It is understood that he built a wing on the Palace in commemoration of the Manila trip . . .

The payment was continued in that sum, notwithstanding the anguished protests of the Sultan from time to time. In 1911 his Majesty visited Singapore, where, after viewing the splendors of the Malaysian potentates, he returned sadly to Sulu to draft an impassioned note to the American authorities. The Sultanic message asked for a pay increase on the ground that the "*influence of the American trusts had greatly increased the general cost of living.*"

The plea fell upon unsympathetic ears.

This then, is the present status of the Sultan of Sulu: salaried man of the Insular government; subsidized man of the British government, and with these things, an appointment as Senator in the Philippine legislative body. Sultan Jamalul Kiram, who passed to the scented gardens of Paradise in 1937, had seldom attended the sessions and had never been known to raise his voice in opinion. His opinions in regard to Philippine independence and the possible general effects upon the circulation of currency would have been interesting.

THE LAST OF THE COTTAS

In the early hours of payday, it was said that the Sultan could be found sitting in an automobile outside the doorway to the government paymaster's office, waiting with a slight degree of regal impatience for his monthly emolument.

But successful financier that he is, it will be as a fighting man that the Moro will best be remembered by the historian. With his cousin, the *pulajan* of Samar and Leyte, he belongs in that select class of men who are born for battle. *Pulajan* and Moro lived according to the law of their land and the code of their country.

Better than anyone else has Stanley Portal-Hyatt summed up this law of the Philippines: "The law of the Bolo has the crowning merit of simplicity. It has but one clause—that the spoil shall go to the man with the longest reach. Possibly the process is crude but at least, it is speedy and final. Judge, jury, counsel: the bolo takes the place of all of these; and there is no appeal, at any rate in this life. The law of the Bolo also has the merit of antiquity. It was in force when the Spaniards annexed the archipelago; it is in force there today; and probably it will still be in force when not only this generation, but half a dozen of its successors as well, have passed away. Not because the law is perfect—no law is, but because it is so admirably suited to local conditions. Half the trouble in the Islands has been because white men do not recognize this elemental code."

It was against that background that Constabulary and Scouts played out that stirring period of American martial history that is represented by the endless patrols.

Chapter XX

SYNTHETIC PEACE

> Officers will endeavor to learn the customs and dialects of the tribes among whom stationed and to gain the confidence of Chiefs and Headmen, so as to be able to exert the influence of the Constabulary in settling disagreements between individuals and bands.
>
> —*Constabulary Manual*

THE combination of rapid Filipinization of the Constabulary and the outbreak of the Great War in Europe was to see the Insular Police losing the services of some of the best names on the roster.

By 1918, the force of regular United States soldiers in the Philippines had been reduced to 8,307, supported by a staff of 7,000 Scouts. These troops held positions in garrison, and were not concerned with the preservation of internal order in the Philippines. The Constabulary, with a force of 5,348 men and 347 officers, became in truth the sole peace agent of the archipelago.

In that year, 88 Constabularymen went to their deaths in the preservation of that synthetic peace. For the Moros were active on all of the southern islands, engaged now in that deadlier form of battle that is represented by small party assaults from the jungle edge.

The Moro outlaws, Sampang and Janni, were eliminated in a series of guerilla fights in Sulu. In Mindanao, isolated *cottas* were still being erected in defiance of the law. When the World War closed, the American participation in the affairs of the jungle police was practically at an end. Coming to the year 1920, but seventeen American officers remained in the corps. The well-known names were gone. Seventy-four men died during that year in the course of the scattered *cotta* fights and the stands against amuck Moros.

The year 1920 saw many small bloody engagements against the Moros. In August of that year vintamen crossed into Dutch territory and murdered a man and outraged two Dutch girls. A Dutch cruiser reported the incident and Constabulary patrols took the field in search of the pirates. The Constabulary launches failed to find trace of the Moros, but a patrol at Bongao, in Tawi-Tawi, captured eight pirates led by Moro Sanwali, who confessed to the foray into Dutch territory.

Lieutenant Abull was killed in a desperate ambuscade in the south, but his men succeeded in eliminating the Moro outlaw, Inda and Pion, who had terrorized the countryside for months.

The Manobos in Cotobato massed near Davao and fought an all-day battle against a Constabulary force. The result was the elimination of the band and the killing of all the leaders. One hundred and two prisoners were taken and sentenced to jail.

In Lanao, Moro Daimla attacked the Constabulary barracks at Ganassi, and was killed leading his bolomen against the police post. During the sacred day known as "Buci Sa Ariraya," a Moro leader named Sarawang gathered fifty men about him and swore on the Koran to avenge the death of Diamla. The avengers took refuge in a *cotta* at Tugaya when a strong Constabulary force under Colonel Waloe and Major Santos appeared. After futile negotiations, the Constabulary took the fortress by storm. Fifty-four Moros were killed.

In 1921, the Constabulary celebrated twenty years of jungle service with a series of bloody campaigns in Lanao and Sulu. In March of that year, Lieutenant Velasquez, with five enlisted men, stopped at the house of the Maharajah Untung in Jolo. They were offered food and shelter, and while sitting at their evening meal were treacherously attacked by a band led by their host. A corporal and one private managed to leap from a window and escape the massacre. Lieutenant Velasquez and three privates attempted to fight their way down a narrow stairway, only to be met at the lower landing by a mob of krismen who chopped them to pieces.

The murderers then withdrew to a fortified *cotta* near Lati, and there an attack was made by Constabulary troops under Captain Nicdao. The engagement lasted from dawn until 11:30 in the morning. When the Constabulary at last were able to enter the silent fort, ninety dead Moros were found within.

The reason for this attack on Lieutenant Velasquez remains unexplained. Maharajah Untung was wealthy and respected by the authorities. Why he should suddenly decide upon a life of outlawry remains an obscure problem of Moro psychology.

Certain it is that the Constabulary lost much of its old prestige and the respect and confidence of the Moro leaders when the corps became predominately Filipino. The Moros were unwilling to accept governmental direction from their hereditary enemies of many centuries. The series of minor combats and disputes that began to appear with regularity could only be traced to the lack of that prestige that went out of the corps when the famous old-timers had resigned.

Men like Crockett and Tiffany and Whitney, Preuss and Furlong and Cochrun, were able in diplomacy and prestige-building, and as capable in the field when necessary. They had the whole-hearted respect of the Mohammedans. A nucleus of these grand old-timers remained; Fort was there, and Bowers and Livingston and Stevens, but their influence was too thinly spread across the long line of Constabulary posts and the lonesome miles of the bush patrols.

Among the high type of native officers in the corps was Captain Antonio Costosa, an old-time fighting man who had seen much service. Costosa had received his education in Spain, returning to the Philippines with an assignment to the Spanish infantry. He then served in the Guardia Civil until that organization was disbanded in 1898.

On March 14, 1920, Captain Costosa was aboard the steamer *Dalupaon,* en route to Tacloban, Leyte. The ship's helmsman suddenly ran amuck. The insane native rushed beneath decks, seized the revolver of Captain Costosa from an unlocked drawer, and began to fire upon the passengers.

Costosa, unarmed, grappled with the maniac, attempting to disarm him. The amuck shot the Constabulary Captain twice in the body and wrenched free. The officer, bleeding from his wounds, went below to search for another weapon and the amuck headed the ship ashore in the direction of a reef. Costosa reappeared with a borrowed revolver and shot the maniac dead.

Antonio Costosa was awarded the Medal of Valor for this act, but he was so weakened by the wounds he sustained that he died in service two years after.

In August, 1921, the Constabulary fought a bloody engagement at Bubung Hill in Lanao, with Lieutenant Nunog demolishing the *cotta* of Amai Binianing after a severe fight. Later in the same month Lieutenant Ballesteros made a thrilling fight for life against Moro Mahang in Parang, near Jolo. Mahang and his band of pirates had held up a Jolo garage truck, murdered the chauffeur and ticket collector, and escaped with revolvers and 100 pesos in currency. After certain treacherous negotiations with the Constabulary on the part of three Moros named Indasan, Israel, and Hassan, in which it was agreed that they should join Mahang and report his movements to the Constabulary, the affair reached a climax on the night of September 24.

On that evening, the three Moros came in to report that they could not find Mahang. In a dark room, Captain Tigno and Lieutenant Ballesteros fought hand to hand with the Moros, who suddenly ran amuck with bared *barongs*. Tigno fell, mortally wounded, and Ballesteros stood with his back to the wall, in the dark, waiting for the attack. A Moro rushed him, and Ballesteros killed him with a stroke of the kris he held. A second Moro loomed up in the darkness slashing at Ballesteros' head. The officer received a slash in the face, and as he reached in turn for the Moro with his blade, the crazed Mohammedan seized him by the throat and both fell to the floor, each trying to cut off the head of the other. The Moro, more expert with a blade, raised it for a killing blow and Ballesteros seized the sharp edge in his hand to prevent decapitation. With terrific fortitude, Ballesteros held that razor sharp blade that cut his hand almost through, hacking all the while at the head and shoulders of the Moro. His enemy slashed to pieces, Ballesteros rounded up the remainder of his patrol before he collapsed from loss of blood.

He is still in service in Zamboanga, survivor of one of the most horrible hand-to-hand battles in Constabulary history. A few weeks later this troublesome band of Moro pirates was rounded up and killed to a man.

In 1923, Lieutenant Angeles survived a severe Moro attack on Pata Island near Jolo. Pata Island is inhabited by the most warlike tribe of Moros to be found in the Sulu Archipelago. For centuries Pata had been a resort for pirates and bandits. It was there that Jikiri had spent much of his time and recruited many of his men. On Pata the Constabulary maintained two small detachments: one at Kiput School House under Angeles, the other at Sapa Malauna under Lieutenant Lasola. In May, because of signs of disorder on Pata, the Provisional Commander of Constabulary combined these two forces into a strong detachment on station at Kiput School House.

The movement of the troops was accomplished with little time to spare, for on May 18 a strong Moro force under Ackbara, Ujaman, and Hatik cut off Lieutenant Angeles's water supply by throwing manure in the well, after which they returned to the swamps to consolidate for an attack on the police detachment. At 3:30 on the next day, the outlaws, numbering some 300 attacked the school from four directions with five columns of warriors armed with spears, *barongs* and *campilanes*. One column, commanded by Moro Isani, succeeded in closing to the walls of the schoolhouse, where they were thrusting with spears at close quarters with the Constabulary. As the attackers massed before the windows, Angeles gave the order for a volley at a range of five feet, and twenty-two of the bandits fell. The remainder withdrew, leaving Angeles in command of the field with no casualties.

But this attack had been but a feeler, and the authorities thoroughly appreciated the gravity of the situation on Pata Island. Colonel Waloe headed an expeditionary force of 112 men and nine officers, and on May 28 a march across the island was undertaken. At Bud Tinga, the village was destroyed, and Upao and fourteen of his followers were killed in a short, savage fight. The expedition then proceeded to Bud Lahi, where the trenches were found deserted. The *cotta* of Mutla was then explored and also found to be deserted, Mutla having been killed in the assault on the schoolhouse.

At the headquarters of the *Datu* Impus, the Constabulary found action. As they came over the hill the sound of the brass gongs was heard, and the exhortations of the priests. Then the voices of the Moros rose in a war song. After two and one half hours of useless parley the advance was ordered. The Moros had steeled themselves to the death battle and no quarter was expected. The running fight was waged against individuals and small bands of amucks who hurled themselves from the walls of the *cottas* and charged the Constabulary force. Thirty-four Moros were killed before the expedition moved on across the island.

The purging of Pata Island ended on a dramatic note. At Niug-Niug it was expected that the Moros would offer the most serious resistance. But the sole occupant was found to be Pandaog, one of the best leaders of the Pata bandits. He had waited patiently there in the *cotta* for the arrival of the troops. As they approached, he raised his *barong* and uttered that shrill war shriek of the Moros, rushing the long Constabulary column with magnificent disdain for death. The Krags roared and Pandaog turned in mid-air, spinning, as the heavy bullets thudded through him. His companions in crime, pushing their war *vintas* through the coral reefs, en route to the main island of Jolo, heard faintly the sound of the volley that killed him.

In 1923, the Constabulary called briefly upon the Navy for support in cleaning out the pirate nests in Sulu. The small gunboats rode down the pirate *vintas* after the Constabulary had flushed them from the jungle. It was tiresome, bitter work, and it accomplished no lasting result. The outlaw bands became smaller, more difficult to find. To cope with the situation, the Constabulary increased their patrol duty. More than 2,000,000 kilometers was the patrol total for the year.

Mindanao came to life again with a serious disturbance in the person of *Datu* Santiago. This Moro came down in force at Bugasan School in Cotobato, and killed three teachers, one Constabulary officer, and nine enlisted men. He was located a few months later in a strong position at Bita Cotta which was assaulted and captured with the killing of fifty-four Moros. Santiago escaped from this battle, to come in and surrender at Camp Kiethley in June, 1924.

The massacre at Bugasan School had been the culmination of many grievances. Moro objection to the payment of the cedula tax had been one cause for rebellion; another had been the forcing of Moro girls to attend Christian schools. These, coupled with proposed taxes on buildings, lands, and dogs had raised the resentment to fever heat. But possibly the principal factor contributing to the unrest had been the increasing uneasiness of the Moros over the growing influence of Christian Filipinos in the Moro country.

This feeling had been plain to the authorities, and Lieutenant Magno had received orders to conduct regular patrols through the affected area. This he had been doing on the morning of October 13, 1923, arriving at Bugasan, where he intended to remain.

The next morning he visited the school, and as noon approached the soldiers sat down for their midday meal. Some thirty Moros were working on the roof of the school building, and *Datu* Gumuyud, a henchman of Santiago's, himself gave the signal for the attack. "What are you waiting for, my children?" he said. "Let us finish our work." Whereupon the Moros working on the roof dropped down among the soldiers and began to cut and slash with *barongs*.

Lieutenant Magno killed two Moros before he was cut down. Sixty Moros engaged the ten soldiers in cramped quarters, and the patrol of an officer and ten men had but one survivor who escaped to Parang to carry the news.

Colonel Waloe placed the punitive party under Captain Fort, and the campaign lasted from October, 1923, until late in February of the following year. Fort whittled away at the band, accounting personally for many of the Moros with his old Winchester "Susan," and in time *Datu* Gumuyud tired of the chase and surrendered. Santiago came in during June of that year, and was sentenced to life imprisonment.

A new coalition of fanatics now appeared to disrupt the peace of the Philippines. Until 1923, the *Colorums,* as they were called, had offered no resistance to authority. In 1918 these religious fanatics had emigrated in great numbers from the islands of Leyte and Samar to Surigao in northern Mindanao, in search of the New Jerusalem. There in the northern Mindanao provinces of Surigao and Agusan they began a gradual withdrawal from the Catholic Church and an affiliation with the Aglipayano Church. This new religion being also not acceptable, they founded the Sacred Family, the sect we know today as *Colorums.*

As *Colorums,* they established headquarters on Surigao Island, and to that place came their Supreme Head, one Laureano Solamo. Affiliated with them was Felix Bernales, who represented himself as the reincarnation of Jose Rizal, returned to life for the purpose of obtaining independence for the Philippines.

These leaders began a campaign of impressing the ignorant *taos* that the world was coming to an end in the year 1924; that the world had slipped so badly in its orbit that it was necessary to collect funds to send to Cebu for a huge hempen rope to tie the earth and keep it from slipping away into outer space. Money came to the leaders in huge amounts.

With these ready funds the *Colorum* leaders began the training of an army of bolomen, destined to take over the government during these last days of the world. Bathing tanks were constructed in the larger communities in which the adherents to the faith bathed to cleanse themselves of all earthly impurities. The tanks were equipped with amulets and holy water and human bones with which the converts anointed themselves to secure immunity from bullets. They were told that if killed in battle, a loyal *Colorum* would be resurrected five days after the battle.

When Lantayog, a new Jose Rizal, arrived in Socorro where a magnificent house and bathing tank had been constructed, the Constabulary moved in to take charge of a serious situation. The town of Socorro was proclaimed the New Jerusalem —the only town that would escape destruction when the earth whirled off into space.

In its capacity as peace and health supervisor, the Constabulary took civil action. A detachment under Captain Juan entered Socorro and destroyed the bathing tank to protect the ignorant people from a disease epidemic. After destroying the tank, Captain Juan toured the country, and then the detachment withdrew with information of the activities of the *Colorums*. Two soldiers were left in Socorro to continue the investigation. They were murdered by the *Colorums*.

Three secret service men were then detailed to worm their way into the fanatical society. Upon landing at Socorro, they were hacked to pieces by the mob. Two more law officers made an effort to penetrate the town peacefully, and they in turn fell before the blades of the fanatics. The hearts of these seven victims were eaten raw by the fanatics.

At this juncture, Captain Juan decided that peaceful overtures were no longer effective, and he, with Lieutenant Guillermo and eighteen Constabulary soldiers, prepared for an armed invasion of the *Colorum* country. The Constabulary launch arrived at the *barrio* of Pamosaingan at four o'clock in the afternoon and found no sign of trouble. After a short wait while the troops were preparing to disembark, one man with a rifle came into a coconut grove near the shore and fired on the launch.

Captain Juan landed immediately. Two detachments—the first, eight men under Lieutenant Guillermo; and the second, ten men under Captain Juan—moved in from two directions upon the silent houses of the *Colorums*. The Constabulary halted when near the houses, and each Constabulary section called three times to the occupants to surrender. At the third call an old man came from one of the houses and shouted to them, "Your guns are useless. You are our victims. Come out, comrades."

The soldiers instantly fired at him and the *Colorum* fell, riddled with bullets. Five others appeared and were similarly killed. Then, a frenzied horde of more than 500 *Colorums* burst from the houses and rushed the small detachment. Captain Juan was killed in the landing boat; Guillermo fell as he gave an order to rally in close order. Fifteen of the soldiers were dispatched in a savage minute of close combat, and the three Constabulary survivors managed to escape to deep water and swim to the launch.

In that short interlude of volley fire that the Constabulary had time to deliver, eighty of the closely packed *Colorums* died. All of the rifles and ammunition of the Constabulary fell into the hands of the *Colorums,* and the immediate result was a rallying of hundreds of recruits to the *Colorum* cause.

This *Colorum* uprising was the most serious obstacle to peace in the Islands that had existed for more than a decade.

The duty of breaking the sect was delegated to Lieutenant Colonel C. H. Bowers, and 500 Constabulary soldiers were sent into the field. With this force was a supporting party of 600 American sailors from the gunboat *Sacramento.*

Colonel Bowers, with 150 men of this force, arrived at Socorro on January 3, and three days later the *Sacramento* dropped anchor off the coral reefs before the town. . . .

The troops were landed immediately, and the guns of the *Sacramento* began to lay down a terrific artillery barrage. The hillsides were combed with shrapnel; the valleys were carefully quartered, and broadsides were laid alongside the town. When the smoke blew away the *Colorums* opened fire vigorously. Pits had been prepared to entrap the Constabulary, but the appearance of the *Sacramento,* with its artillery, had changed the plans of the *Colorums,* who were set to repel an infantry attack. A slashing defeat was the result here, and it was followed by a similar strafing of the mainland town of Timamana.

As these battles were in progress, information was received that the *Colorums* were threatening the town of Placer on the mainland of Surigao; and at Bad-as, seven miles from Timamana, 200 *Colorums* were organized to resist all law. A Constabulary column moved in, and was immediately charged by the *Colorum* force. These fanatics were able successors to the deadly *pulajans* of Samar, for they crawled, riddled with bullets, until life left them. Some of them struggled to their feet in a last effort, and hurled their bolos at the troops they opposed. Fifty-four of the *Colorums* were killed at Bad-as, and the remainder were wounded or captured.

Then, on February 12, 1924, the last massed *Colorum* attack occurred in Agusan. Some hundred survivors of the bombardment at Socorro ambushed and killed Lieutenant Mendez and ten of his men in a narrow trail near Balite-

Tibung. Thirty-five *Colorums,* including the two leaders, fell in that affray before the Constabulary were cut down to a man.

The ambush of Lieutenant Mendez finished the active resistance of the *Colorums* in Mindanao. The sect was to reorganize again, in Luzon, the succeeding year.

Chapter xxi

THE END OF THE CORPS

> "President Manuel Quezon today signed a proclamation transferring the Constabulary to the army and increasing the force from 5,500 to 9,000 men . . ."
>
> —*Newspaper clipping*

THE last decade.

In 1926, the Constabulary was conducting the usual field operations in Mindanao and Sulu. The force consisted of 132 companies, with an artillery platoon, and they were in garrison in 162 stations in the archipelago. 6,053 men, they had; and only twenty-five of the officers were Americans.

A reorganization had come into the geographical districts. There were but four divisions now: the Department of the Visayas; the Department of Northern Luzon, and Department of Southern Luzon, and the Department of Mindanao and Sulu.

Lieutenant-Colonel Luther R. Stevens had the hot spot in the Moro country; he had a busy year in 1926. In May he was in the field in Lanao, in operations against the Moros at Tugaya. Six *cottas* were captured and seventy Moros were killed. The Constabulary lost six men killed and four officers and thirty-two men wounded.

In September, the patrols were on the island of Tawi-Tawi, in the tangled woods back of Bongao. A band of Moros was discovered in a sudden jungle that was swept with rain which made visibility uncertain. It was the old story of standing to the rush of wild krismen. The Constabulary held the field, to break even with the Moros. Eight Mohammedans died in that yelling frantic attack; three Constabulary soldiers died with them, and five others of the police were seriously wounded.

During that year, the force had sixteen encounters with outlaws. The records show the capture of twelve Moros, the wounding of ten and the killing of ninety-six. The ferocity of the resistance set up by the Mohammedans is shown by the casualty lists. In inflicting 109 casualties upon the Moros in sixteen engagements, the Constabulary suffered a loss, in killed and wounded, of 116 men and officers.

The Moros made it man for man. Here again it was demonstrated that the kris, at close quarters, has not been improved upon by the modern armaments of man.

The years rolled along and the battle mood persisted. There was no let-up from the constant Mindanao and Sulu patrol. Red-epauletted men still struggled for the peace of the Islands in a day that was only yesterday. In 1927, the grand old Chief, Rafael Crami, passed away. He was replaced by C. E. Nathorst, a doughty Norseman, who took up the command in a bad year of internal disorder. The middle islands came to life again, after a long period of peace.

The Emperor Flor Intrencherado, one Florencio Nativdad, caused a brief flurry of unrest in Negros. Intrencherado had been a valuable aide to Captain W. A. Smith of the Constabulary years before, in the pursuit of "Papa" Isio. Nothing was heard of the man for almost two decades until he suddenly came into prominence in Jaro by raising a red flag above his house, and beginning to preach of his supernatural powers.

A fortunate venture into the dried fish business in 1921

netted him the sum of 11,000 pesos, and with that he returned to Jaro to lay plans for an ambitious future. In 1925 he began a campaign for his "Imperial Government." At this time he announced his candidacy for the Governorship of Ilo-Ilo with a platform committed to the division of the island of Panay into several dukedoms with himself as "Emperor of the Archipelago."

He was taken in by the Constabulary on several occasions, serving short terms in jail, but we hear of him late in 1925 claiming to be in communication with a Spiritual Guide and threatening to bring down upon the people floods, pestilence, and famine if they refused to follow him.

He was declared insane, but natural catastrophies that followed increased his prestige fivefold. For one thing, Canloan volcano in Negros erupted; there was an earthquake in Japan, and a civil war in Mexico and China. The ignorant, remembering his prophecies, flocked to his standard. When they reached the total of more than 26,000, Intrencherado decided it was time to take over the government. Consolidating 300 of his men at Victorias Negros early on the morning of May 13, 1936, he instructed them to take over the town. Intrencherado was not there in person; he sent as his Lieutenant one Policarpo Montarde, who wrapped a white towel about his head and entered the Municipal building, stating he wished to read a "Law from Paris."

No officials being present, he attacked the police force, killing two patrolmen before a Constabulary detail under Lieutenants Ruffy and Cortes arrived and restored order.

Meanwhile, at Jaro, a squad of eight Constabulary with one officer was guarding the home of Intrencherado, prepared to forestall any further outbreaks. Four hundred men gathered to call the Emperor out to lead them but the Constabulary stood firm, and the "Emperor of the Archipelago," hopelessly insane, was removed to San Lazaro Hospital in Manila, where he remains today.

Then Nathorst turned back to the weary task of subduing the Moros. In January, the corps was in Sulu, before the *cottas* of *Datu* Tahil. *Lantakas* flared and men died before Tahil was captured and his hilltop fortress razed to the ground. The same month, the Alankats, ferocious mountain Manobos of Cotabato, flared in revolt under the leadership of Angkay and Maon. Colonel Stevens was in the field—this lanky, red-haired Mississippian, so thin that his associates would say to him as he left for patrol duty, "Turn sideways to them, Steve: no Moro could hit the edge of a knife blade."

Stevens knows Mindanao and Sulu as is the doubtful privilege of few white men. For a decade, he served as Chief of the Department of Mindanao and Sulu; he served the Insular government well. Today, he is among the last of the American old-timers and is in the service of the new Philippine army.

Certainly one of the most picturesque as well as the most dangerous figures in Moroland was Mampuroc, who had instigated the Alangkat movement among the Manobos in Mindanao. This was one of the most formidable native combinations of the later years, and its effects would have been far-reaching had not the Constabulary taken prompt and decisive action.

Mampuroc was probably the greatest Moro figure since Jikiri, and his short career of banditry had inspired legends and stories that were told about the Manobo fires in the mountains.

It was said that Mampuroc was immortal and beyond the reach of mortal punishment. He was, it was said, the reincarnation of *Datu* Alik, that famous Moro who had ruled Mindanao in an earlier day. Others of his tribesmen followers believed him to be the Spirit of Mohammed. And to the most rabid of his men he was "The God on Earth."

So great was the influence of this demi-god that the trails approaching his headquarters at Vintigan were blocked and choked with hordes of pilgrims bearing him gifts of coconut oil, red cloth, and edged weapons.

Along these twisting, winding trails the Constabulary patrols hastened, desperately seeking to cut off this menace before Mampuroc's followers had grown to the thousands in number.

Stevens made every effort to settle the affair without bloodshed, but Mampuroc was in no mood to treat with the government. He refused to meet the Colonel or the Governor of the province, and told his followers that Stevens and the Governor desired to become Alangkats.

Mampuroc was a man of great intelligence and great personal magnetism. He was able to convince his wild followers that he was indeed a god, and he began a systematic program of steeling them to face modern weapons of warfare. It was established as a fact in his camp that his magic wand (an umbrella stave mounted on an empty cartridge case), would make attacking soldiers blind; that a secondary black coral wand would turn bullets into rain; and that his wooden crocodiles would come to life and soar into the sky to combat enemy aircraft.

The followers of Mampuroc were among the most dangerous tribes of the Islands. The Manobos are stocky, sullen bowmen of Mindanao. They are equally facile with the kris and the *barong*. They wander a wild territory in Southern Mindanao; their fires high in the hills can be seen by the travelers who come down across Mindanao on the *cogon*-lined jungle trails. Technically they are not Moro, although many have adopted the religion of Islam.

With this raw, wild material at hand, Mampuroc began to develop an army. As the recruits flocked to his headquarters, he carefully disarmed them, rearming only a chosen few. After each new arrival had passed through a probationary period, he was armed in turn. The result of this stroke of genius was a frantic scramble for favor and a great disciplinary effect on the hillmen. The bearing of arms became a privilege that could come only after devotion to the god.

In many ways, Mampuroc was the most sincere leader the tribes of Mindanao were to produce. He taught his men to work; he forbade stealing; he frowned on marital intimacies between man and wife without the full desire of the woman. He was a strong leader, and he was rapidly welding a compact, loyal following when the Constabulary came. Regardless of Mampuroc's revolutionary tendencies, impartial history of the region will record him as a sincere leader. He disapproved of the payment of cedula tax, of sending children to the Christian schools, and of organized government in general.

In 1500, he would have been the ruler of all the Philippines. In 1927, he was a government menace—and he was removed. A strong force of Constabulary invaded his territory, and on the morning of March 23, 1937, Mampuroc and thirty of his followers met their deaths. The Alangkat movement collapsed with the death of the Messiah, although patrols in the region were necessary for several weeks until the natives had lost their conviction that Mampuroc had escaped, or that he would return from the dead.

Actually, Mampuroc was found 200 meters from his house with five Springfield bullets in his chest, and he was buried in the jungle where he fell.

Combat against Moro *cottas;* pursuit of cattle thieves; the quelling of mutinies and public brawls—all were a part of the duties of the tropic police almost within the decade in which we live. In 1927, a Medal of Valor was won by a youngster en route to his first patrol.

It was on the steamship *San Antonio* near Oroquieta, Misamis, on January 30. Third Lieutenant Robert Young was a passenger on the ship; he was on his way to Dansalan, Lanao, to fight Moros. At eleven o'clock in the morning another passenger worked himself into that sudden murderous rage that is the habit of the Malay. The amuck seized a shotgun and fired at the Captain of the ship. As the Captain leaped to his cabin, the maniac killed the two helmsmen, and began to attack the passengers with a knife. Lieutenant Young was unarmed, but he was an officer of Constabulary. He hurried across the deck, ordering the crazed attacker to lay down his weapons. The shotgun spouted flame and Young was seriously wounded in the arms and legs. Young turned away and started below to his cabin to secure a revolver. In his absence the killer slew three of the passengers, and began to lower a boat to make his escape.

The Constabulary officer had no time to secure a weapon; he hurried back, unarmed, and grappled with the amuck native. He was stabbed to the heart and killed.

This young Lieutenant learned early the traditions of the men with red epaulets; he had been a Constabulary officer for 107 days, and was en route to his first assignment.

In 1928, Stimson, the aloof and courteous, was enjoying a brief tenure of office as Governor-General. Independence talk was in the air—but there were fifteen encounters with Moros in Sulu and Mindanao. Men were still dying for the peace. 1929 was quiet—but six Constabulary gave their lives in the nine encounters with Moro outlaws.

In 1930, a Moro Sultan, one Mamur, began to develop a following in the dread Lake Lanao country that may never be quieted without mass extermination of the Mohammedans. Mamur built *cottas* of heavy posts and hard packed earth, with bamboo thickets to protect the walls. By devious means, he acquired unlicensed firearms. He was ready for open rebellion.

Lieutenant Mayne took a detail into Lanao, with orders to bring in the unlicensed arms—or Sultan Mamur. He did neither; he died on the *cotta* walls in a magnificent, bloody assault that ranks with great fighting epics of the glamorous days of 1906. With Lieutenant Mayne that day, died four of his men; and six officers and thirteen privates of Constabulary were wounded in that capture of a Moro *cotta* and the killing of nine of its defenders.

The encounter was a desperate charge across open country to the walls of the flaming *cotta,* which was defended by brass *lantakas* and a swarm of krismen.

Twenty days later, on May 28, Captain James R. Grimstead led a Constabulary force against the *cotta* of *Datu* Gnassi at Tugao, Lanao. The fighting was a horrible hand-to-hand combat on the walls.

As one reads of these encounters, and visualizes that terrific resistance of the Moros and their quite evident ability to inflict casualty for casualty, there comes to mind again that often discussed question of the relative merits of *pulajan* and Moro as a fighting man.

All of the Malays are valiant; there can be no doubt of that; but of the two, in the opinion of the writer, the palm for fighting genius must be awarded to the Mohammedan of the south. They have had more practice, for one thing. The *pulajan* resistance lasted for a short decade; the Moros have raised a kris against every nation in the world, almost, in that magnificent defense of their island empire. Their battles are measured by centuries of time.

Even in the face of such able summaries of *pulajan* and Moro as have been made by Captains Preuss and Holmes, this statement of Moro superiority must stand. Preuss and Holmes fought both *pulajan* and Moro, in the days when they had to send two men with the cook as he went to dig *camote* roots at the edge of the stockade walls. The two Captains stated that the hill Filipinos *(pulajans)* fought equally as well as the Moros—*if and when they had as good anting-antings.*

In that qualifying clause rests the superiority of the Moro. No *anting-anting* of holy oil or scrap of paper scrawled with the name of a "Pope" could equal the religious *anting-anting* —the Koran of Mohammed— which guaranteed repose with the virgins in scented gardens to the men who died in battle.

In the Mohammedan country, the Constabulary faced an antagonist who fought under wily jungle rules of his own making. Here were no *pulajans,* rushing in massed formation to death; here was a calculating Moro, who used every spear of grass for concealment and who had developed homicide into a science. The Moro was as willing to die as the *pulajan;* he proved that on many occasions. But in that pleasant transition to Paradise, the Mohammedan expected to carry with him certain Christians who were in his last line of attack on earth. Almost always, the Moro rode to Allah on his white horse, with the shades of two or three Christians at his saddlebow.

The Moro is harder to stop than the *pulajan*. The *pulajan* can be killed. The Moro can be killed too, but the process is infinitely longer and more complicated. Furthermore, the Moro requires an unbelievable amount of lead to accomplish that desirable result. He absorbs bullets without seeming effect. He carries fanaticism to a depth unplumbed by the *pulajan;* almost, the Moro carries his fanaticism to the point where it defies death itself.

The last decade rolled along; 1930 ended and the time came for a summation of the activity of a year. Again the Moros had been consistent in that matter of inflicting casualties. Fifteen Moros had died at the hands of the patrols in seven encounters. One officer and four privates of Constabulary had died with them, and six officers and fifteen men had been wounded.

1931 . . . two officers and three men died during an assault of the Constabulary barracks at Tayig. Captain Leon Angeles met his death in Jolo; jungle clearings still ran red with the blood of men.

As early as 1925 there had been indications that peace conditions in Pangasinan and Nueva Ecija were threatened by another powerful fanatical coalition. An organization known as "Kapisanan-Kabola-Kasinag," numbering some hundred members, had appeared in these provinces. Among the most active leaders was one Juan Lago, who set himself up as St. John the Baptist, second in command to General Cabula.

This Cabula, the organizer, was well known to the forces of law and order. In Pangasinan he had been accused of swindling and sedition, and in Nueva Ecija of sedition and rape. The organization he built was military in character, the members having commissioned and noncommissioned ranks, and wearing a prescribed uniform of red blouse, blue trousers, and red hat. The aim of the society was the so-called betterment of the society members by a division of the land and of personal property among the poor.

Revolution was to obtain this desired result, and the plans

called for an attack on San Jose on Friday, March 13, 1925, after which a new government would be set up under General Cabula.

As a means of identifying members of the "K-K-K," injections of a mercury solution were made in both arms and in the thighs, under the belief that such injection gave great physical strength and an immunity from bullets and the bite of poisonous reptiles. Each member paid a three peso initiation fee, and a total of eighty centavos for the injections. As a result of the injection of the mercury compound, a knot rose on the skin that remained with the members until death.

When the membership of the organization had reached a total of 2,000, Cabula ordered an attack on the Constabulary. On March 3, 1925, the misguided fanatics moved against the Constabulary station, and in the battle that followed, General Cabula and five of his men were killed. The Constabulary had one private wounded by a spear thrust.

The society was six years in recovering from the loss of their leader, but in the early morning of January 11, 1931, a band of sixty K-K-K members made an attack on the Constabulary station at Tayug in Pangasinan. Moving stealthily, at two o'clock in the morning they completely surprised the Constabulary station, silencing the sentry with bolos before he could give the alarm. The barracks were burned and all the arms of the detachment carried away. With the town defenseless, the fanatics then burned the Municipal Building and the Post Office, and sacked the town of Tayug. One building they spared — the Catholic convent. Inside its thick walls they took refuge and made provision for a stand. Reinforcements of Constabulary came immediately to the scene, and after a fierce two-hour battle the convent was captured by the Insular Police.

With that defeat, this flaring of the *Colorum* movement

in the north died away, leaving the Constabulary free to turn their entire attention to the Moros.

On October 9, 1932, near Camp Seit, in Jolo, a Constabulary patrol under Lieutenant Vicente Alagar was rushed by hostiles. Alagar and thirteen of his men were killed; the nine survivors extricated themselves from the scene with difficulty. *Imam* Ibra exhorted his followers to great heights of fanaticism that day—the kris blades bit deeply during that close range ambush.

Then on September 6, 1933, Lieutenant Julio F. Barbajera was leading a patrol on Jolo Island. The whistle of a spear was the first warning of attack, and Barbajera went down, his detail rushed by a few Moros. The Mohammedans, under Mahamud, had a field day from ambush at the expense of the Constabulary patrol. With Barbajera, the Police left six privates dead on the field.

On February 9, 1934, Lieutenant Barbajera received a posthumous award of the Medal of Valor.

Sixty days later, on November 20, Lieutenant Mariano G. Esculto, recently appointed commander of Camp Andres in Sulu, took six men to investigate a murder near the Constabulary station. He was striding through a clearing at the head of his soldiers, when the party walked into ambush. Esculto fell at the first shot from the bush, and his six men were able to drop the three Moros who launched an attack with kris and *campilane* blades.

Ambush, with men skewered on spears—this in 1934!

On November 14, 1935, the last Medal of Valor was awarded to a Constabulary soldier. Seventeen years after he had led a detachment across a swaying ladder at Bayang *cotta* in 1917, Paulino Santos was decorated with the highest award of the corps.

That day in 1917, a high *cotta* wall had required the

placement of scaling ladders before the assault could be made. The American officer commanding the detachment called for volunteers. All four of the American junior officers stepped forward as one man, but before a decision could be made a tall young Filipino spoke, "Sir, I should be allowed to lead this assault. They say Filipinos cannot fight Moros. Here is the opportunity for a Filipino to lead Filipinos in the assault of a Moro *cotta*. I should be allowed the command of the assault party."

And so he was awarded leadership of that desperate party. He placed the ladders and led the way to the crest of the walls. Every Constabulary soldier in the attack was wounded and the *cotta* fell.

Today, Santos is Major General and Chief of Staff of the new Philippine Army.

One of the last operations of the Philippine Constabulary as a unit of fighting men—this one, a brief and bloody encounter in the new Province of Bukidnon—in Mindanao. It is narrated in a clipping from the *New York Times,* a brief news report, with no hint of drama. The editor was unaware of the fact that he was practically writing the epilogue to a grand career of jungle service. This is combat in 1936:

> "One Constabulary man and four bandits were killed today in a clash in Bukidnon Province, Constabulary headquarters was advised today. The fighting started when the soldiers called to the bandits to surrender and were answered with a volley of shots. Fabello, the gang leader, was killed, with three of his followers. The rest surrendered."

In that account is one significant phrase: "The soldiers called upon the bandits to surrender." Always, the antagonist had the right of the first shot. The Constabulary was the agent of law and order; they killed to preserve the peace.

Jungle Patrol

THE END OF THE CORPS

On January 14, 1936, President Manuel Quezon, of the new Philippine Commonwealth, sounded the end of one of the most storied bodies of fighting men in the history of conquest and war. His pen signed an order transferring the Philippine Constabulary to the new National Army of the Philippines. The collar insignia, "P. C.," gave way, after thirty-six years of jungle service, to a regular military establishment of regiments and brigades.

In summing up the campaigns of the Philippine Constabulary, a discussion of the weapons at hand or the marksmanship of the men is not sufficient to explain the greatness of these jungle campaigners. The point involved is their terrain of battle.

The rifle and the revolver and even the machine gun lose much of their authority in dense jungle. The visibility is poor and the firing range exceedingly short. The number of rounds a man can fire is limited; too quickly, the combat reaches close quarters. With a Krag rifle and a .45 Colt revolver, every Constabulary soldier of the later days had a potential firing possibility of eleven shots without reloading. He was often outnumbered twenty to one, or more than twenty to one. The principle of the campaigns involved, not the destructive possibilities of the eleven shots at his command; to be considered most was that grim element of time. In the face of a sudden bolo rush, the police often had time, for but two or three shots before the action was man to man. And against impossible odds.

For bruising shoulder-to-shoulder work, the native weapons remain the best in that jungle scene that developed them. At close quarters, the Moro kris or the *pulajan talibong* have destructive qualities not surpassed by the modern automatic pistol or the sub-machine gun.

THE END OF THE CORPS

The passage of a high-velocity bullet through the body is killing but not immediately fatal; sometimes, in the heat of battle, men can remain on their feet, desperately wounded, for a lengthy period of time. But the last despairing swing of a bolo blade, in the hands of a dead man riddled with bullets, could be deadly. And the blow of an edged weapon has a finality about it; it knocks a man out of action quickly. Too, there is a mental menace in the facing of bright razor-edged steel.

Upon these facts of the nature of the native resistence, the men of the Constabulary may rest their greatness as fighting men. Against the *pulajans* and Moros of 1904 and 1905 and 1906, a company of machine gunners would have had horrible moments of fear and doubt as they were rushed by a thousand fanatics in close and crowded quarters.

The work of the Constabulary, in facing that jungle horde with single-shot weapons, is little short of miraculous. Battle in those days required a fiber that is not necessary in these days of casual long-range work with the enemy not often in sight. In that Philippine jungle the Constabulary developed a degree of bush knowledge, superb judgment, and fighting genius that stamps them as among the greatest individual fighters of history.

They rest on their unvarnished combat reports.

The officers of the Philippine Constabulary were consistent men of battle. To many of them, the conquest of the Philippines was but a facet of a lifetime profession at arms. Crockett, Rivers, Harbord, White, the two Griffiths, Preuss, Allen, Bandholtz, and many others saw service in France.

The rapid fighting pace of the Constabulary had eminently fitted them for future responsibilities. Nor did that responsibility weigh with undue heaviness upon their shoulders; they had taken great responsibility, twenty years earlier, while youngsters in the Philippine Constabulary.

Some of the later services of these officers were gilded with an aura of great romance. They scurried away to settle boundary disputes and to serve as military attachés in distant ports half a world away. They were living characters from adventure books. There was never doubt or indecision where they were; they would have been the first to scoff at the suggestion that they were figures of romance. E. R. Griffiths was one of them the—first American officer to fall in France. We can imagine him there, hurrying with that restless Constabulary pace to the front line to die. Across a political boundary line, in Flanders, his fellow Constabulary veteran, R. H. Griffiths, met the last great adventure during that miserable retreat of the British army from Mons to Ypres. Two valiants walking hand in hand out of life, in battle, as became ex-Constabulary men.

We have an intimate picture of Bandholtz, suffering great disappointment in the World War. For a while, he had been at the front, proud to command his brigade in action.

Then General Pershing sent for him. Bandholtz himself told of that interview that turned an old war-horse away from the bloody fields of France.

"Now, Bandholtz," Pershing said, "you are going to hate me for this. But the Provost Department is in a disgraceful condition. I want you to take hold of it and put it in shape. When you can come to me and say that the provost guard is working to *your* satisfaction, you can go back to your command at the front."

So Bandholtz took over the 22,000 officers and men serving on provost duty in all of the allied countries, including military police jails. It was a tremendous and a dirty job. The war ended with Bandholtz Provost Marshall.

And then, long after the war, a statue was erected, in Budapest, Hungary, to an American general. A mob had been bent on wrecking the government building, which was without police protection. It had seemed that the mob would have its way, when an American army officer stepped out, armed with a riding whip and a determined manner, and stopped them in their tracks and sent them away.

So there in Budapest today is a statue of General Bandholtz with his riding crop, flinging a gesture to the past. To a past that is tied up, more closely than anyone could possibly know, with the capture of a wrinkled little brown bandit named Simeon Ola, in the jungles of the Philippines.

It was the old Constabulary gesture of men against odds.

And so Bandholtz came to the end of a life that had been magnificently full. He had returned to his home at Constantine, Michigan. His wife came home one afternoon to find Bandholtz peacefully dead, sitting in his garden.

With Garwood the glamorous, we bring this account to a close. His stride through life was a romantic gesture. Garwood, the great ladies man—"How, Madame, can a beautiful and charming lady like yourself, throw herself away on a shriveled-up little cuss like your husband?" Gesturing beneath General Allen's nose with his long cigar; swaggering across the Philippine Archipelago with his hundred pairs of shoes and his two pistols that he could use so well; riding into ambush and mountain conference and steaming swamp —and loving it all.

Gunfire and glamour and fair ladies and Garwood. He had a facility for finding trouble. His Constabulary days finished, he joined the famed Pennsylvania State Police as one of the original members.

"No," they told him once, "you can't arrest that man. He's in a saloon, surrounded by his armed mob."

"Can't I?" said Garwood. "A policeman can arrest anybody—anytime. If he can't do that, he isn't a policeman."

And then, as a Lieutenant of Pennsylvania State Police, he resigned to wander away again, with those bright eyes fixed on new and fascinating horizons. He escaped the bandit bullets as he had made light of the *pulajan* blades in the Philippines. It was not written that this man of action should die in battle.

New experiences; new scenes; new friends. Garwood could never grow lonely, nor old, nor sated with life—he could draw at will from his store of magnificent memories. He died peacefully, as so many grand fighting men do; in a hospital bed at Omaha, he jested with the surgeons as his life ebbed away.

He was never completely serious; he was always restless; and never, in the slightest degree, did he understand the meaning of that grim emotion that is fear.

It was Jesse S. Garwood most of all, possibly, who best typified the spirit of those dauntless, dreadful days when men of the Philippine Constabulary were engaged in the conquest of jungle.

The jungles of Samar and Mindanao still remain—to be pointed out as the symbol of an island empire that has felt too lightly the impress of civilization. America is laying aside "the white man's burden" that was the motive force of the Constabulary gallants of three decades in the past.

The Philippines are in process of abandonment.

THE END OF THE CORPS

Today, on the edges of that mighty jungle, the curious may possibly find in Samar a rounded hill that was once a *pulajan* fortress—or in Mindanao, a crumbling ruin that once flamed with *lantakas* as a *cotta* of the Moros. As the curious one gazes at these evidences of a stern day, there may come a brief awareness of the romantic martial history that broods over the solemn bush.

The tourist may look about him and see here, a place which once sounded with the dreadful patter of bare feet that signaled the rush of the bolomen; or there, a place where a lone Constabularyman won a Medal of Valor in 1906. But only the most imaginative will be able to reconstruct those jungle battles that are thirty years in the past.

The men who haunted that bush—and made it forever their own by right of conflict—are forgotten by a nation that forgets too easily in the press of other, greater wars.

Only the jungle remains—waiting to be forced again by some lesser breed of men.

The End

Bibliography

DURING the search for material the author became aware of the fact that the general subject of the Philippine Constabulary had been sadly neglected by the writers of American battle memoirs. There was in existence no study of the Philippine Constabulary, and very little on the corps in the books devoted to the Philippine Islands.

During the research, the following volumes came to light, all of which have been used in comparative checks to make the material herein as accurate as was possible.

Bullets and Bolos, by Colonel John R. White, is an excellent account of the author's personal experience on Negros and Mindanao. Colonel White has a flair for narrative, and his book is rich in the color of the day.

The Philippines, by Charles B. Elliot, has a brief mention of the organization of the Scouts and Constabulary, and a few footnotes on representative outlaw chiefs.

With this limited material available it became apparent that any account of the Philippine Constabulary must be the result of contact with individuals who had been present during those stirring days of its inception. A systematic attempt was made, therefore, to get into personal correspondence with retired officers, and to compile, from their personal accounts, the data concerning the corps. As a result of these personal interviews and a voluminous correspondence, the material was gradually accumulated over a period of fifteen months.

I should mention also the slender volume entitled *Medal of Valor,* by Major E. Baja, of Manila, in which is detailed a reprint of the General Orders awarding this medal to the individuals of the force who were so honored. Major Baja has made this compilation without comment, contenting himself with the setting down of the official orders as they emerged from headquarters.

Colonel Cromwell Stacey, Colonel Cary Crockett, and Captain A. E. Hendryx and Captain Jesse Tiffany contributed valuable combat material and much time to help in the preparation of this volume. I would have been unable to finish it without their enthusiasm and support.

In presenting these notes on the Philippine Constabulary, I do so feeling that the book belongs to the forty-odd officers who so generously co-operated with me in furnishing the material. I feel also that this stirring period of American martial history has been shamefully neglected. The characters who undertook the subjugation of the Philippines have received so little, in honor or in financial award, for the consummation of that very dismal business.

Therefore, blanket credit to the men who thumbed the leaves of their personal experiences to turn to me their personal records of those bloody days of the first decade of the century.

I wish to thank Mr. Perry Hiskin, a busy purser of the Dollar Line, for material he brought to me in the course of his various trips across the Pacific. And to Norrie Miles, of Manila, my thanks for the care with which he has attended to the numerous questions I have forwarded him. He spent much of his personal time in seeking the answers to questions that arose during the writing of the book.

In a volume detailing the frenzied activity of a force of jungle police over a period of almost forty years, there are certain to be errors and omissions. Many men worthy of pages in this volume have been misplaced by the years that have flowed to mist their activities. I have selected a campaign here and there, as representative of the jungle wars, in an attempt to bring to the reader a cross section of the bush and of the men who formed the background of the times. This is not so much a history of the Philippine Constabulary, as it is that cross-section picture.

Possibly some of the incidents have been slightly misdated. I have spared no trouble in the effort to keep the text accurate, but the reader should remember that this volume was written mostly from personal memoirs, and minds do not always agree on the relatively unimportant matter of "when." In all cases where the contributors failed to agree upon the date of an identical incident described by all, I have referred to the reports of the Chief of Constabulary and have taken the date therein specified. It was quite the common thing to pursue a bandit through pages of personal letters, notes of interviews, and printed combat reports, to find him respectively killed on Samar Island; captured and sent to Bilibid; hanged at Surigao; or escaping, scot-free, into the bush. In such cases, the Chief of Constabulary has been the argument settler.

In the main, I stand on the text as presented.

Appendix

THE casualty lists had become a matter of concern in the Constabulary, as they had no such provision for retirement and disability pay as prevailed in the Army. Realizing this, a pension fund had been established some years after the organization of the Corps which provided for an internal fund for use in such cases. Pay deductions were made from the various ranks as follows:

Colonel	Deduction from monthly pay—$1.75
Lieutenant Colonel	Deduction from monthly pay—$1.50
Major	Deduction from monthly pay—$1.25
Captain	Deduction from monthly pay—$1.00
First Lieutenant	Deduction from monthly pay—$ .75
Second Lieutenant	Deduction from monthly pay—$ .75
Third Lieutenant	Deduction from monthly pay—$ .50
Enlisted man	Deduction from monthly pay—$ .10

These monthly deductions were set up into a retirement and disability fund for distribution to the retired and their dependents.

Disability classifications were established as follows: Loss or loss of use of both hands, both feet, both legs, both eyes, or any two limbs—total disability. Loss of one arm, one foot, one hand, one eye, all fingers of one hand, or loss of hearing — ⅔ disability. Loss of one eye, or two or three fingers — ⅓ disability. Retirement was provided for at the age of 55, after 20 years service, with a maximum of 75 per cent of base pay. The widows of Constabulary soldiers dead in line of duty were to receive ⅔ pay.

The pitifully small retirement pay provided for by this plan is indicated by the following total disability tables:

Brigadier-General

15 years service and less than 20	—Pension $40.00 monthly
10 years service and less than 15	—Pension $35.00 monthly
5 years service and less than 10	—Pension $32.00 monthly
Less than 5 years service	—Pension $30.00 monthly

Colonels and Lieutenant-Colonels

15 years service and less than 20—Pension $35.00 monthly
10 years service and less than 15—Pension $30.00 monthly
5 years service and less than 10—Pension $27.50 monthly
Less than 5 years service —Pension $25.00 monthly

Majors

15 years service and less than 20—Pension $30.00 monthly
10 years service and less than 15—Pension $25.00 monthly
5 years service and less than 10—Pension $22.00 monthly
Less than 5 years service —Pension $20.00 monthly

Captains

15 years service and less than 20—Pension $25.00 monthly
10 years service and less than 15—Pension $22.50 monthly
5 years service and less than 10—Pension $20.00 monthly
Less than 5 years service —Pension $17.50 monthly

1st Lieutenants

15 years service and less than 20—Pension $20.00 monthly
10 years service and less than 15—Pension $17.50 monthly
5 years service and less than 10—Pension $15.00 monthly
Less than 5 years service —Pension $12.50 monthly

2nd Lieutenants

15 years service and less than 20—Pension $15.00 monthly
10 years service and less than 15—Pension $12.50 monthly
5 years service and less than 10—Pension $11.25 monthly
Less than 5 years service —Pension $10.00 monthly

As may be seen from the above, a Captain with twenty years' service, retired for total disability, would receive a monthly pension of $25.00. There *was* no retirement from the Constabulary service. The ex-officers who survive today are employed in civilian activities. The governmental efforts made to reward these muddy riflemen who made the Philippines safe are conspicuously absent and remain our national disgrace.

One sop was provided, but that only on special order of the Governor-General after specific investigation of each case. That was the privilege of the Governor-General, in special cases, to retire Constabulary soldiers as follows:

Generals . . . sliding scale, according to length of service, to receive monthly amounts ranging from $75.00 monthly to $112.50 monthly for a period of *Five Years* after discharge.

Colonels . . . sliding scale, according to length of service, to receive monthly amounts ranging from $62.50 monthly to $100.00 monthly for a period of *Five Years* after discharge.

Majors . . . sliding scale, according to length of service, to receive monthly amounts ranging from $50.00 monthly to $85.00 monthly for a period of *Five Years* after discharge.

Captains . . . sliding scale, according to length of service, to receive monthly amounts ranging from $45.00 monthly to $75.00 monthly for a period of *Five Years* after discharge.

1st Lieutenant . . . sliding scale, according to length of service, to receive monthly amounts ranging from $40.00 monthly to $62.50 monthly for a period of *Five Years* after discharge.

2nd Lieutenant . . . sliding scale, according to length of service, to receive monthly amounts ranging from $35.00 monthly to $50.00 monthly for a period of *Five Years* after discharge.

Enlisted men . . . sliding scale, according to length of service, to receive monthly amounts ranging from $4.00 monthly to $10.00 monthly for a period of *Five Years* after discharge.

General J. G. Harbord, who saw all sides of the service in the Philippines, has ably summed up the treatment of the Constabulary in a foreword to the manuscript of Jesse A. Tiffany. He says, "The romance and adventure of the Constabulary service, particularly in the Moro country, would furnish the theme for a score of Kiplings, Remingtons, or Wisters. The officers who lived the adventure are scattered across the world. Some of them fell in France when America entered the World War. A very few, perhaps a half-dozen, are still in service in the Philippines. Others who had entered the Constabulary as a career could not foresee the fate which overtakes the servant of Colonies,—and have given away to native officers. Out of step with affairs at home, and weakened by a long stay in the tropics, they exemplify the ingratitude of their native land. Our country has not appeared to be interested in their fate."

JUNGLE PATROL

Index

JUNGLE PATROL

Index

JUNGLE PATROL

Index

Index

Index

Index

Index

Index

Index

CPSIA information can be obtained at www.ICGtesting.com
Printed in the USA
LVOW06s0946080913

351490LV00001B/88/P

9 780983 475620